MySQL 8.0
从入门到实战

张文亮 编著

清华大学出版社

北 京

内 容 简 介

MySQL 数据库是目前全球流行的数据库之一。本书从入门到实战，系统全面、由浅入深地介绍 MySQL 数据库应用的各个方面。

全书分为 8 个部分，共 18 章。第 1 部分（第 1~3 章）介绍 MySQL 的基础知识，包括初识 MySQL、数据库设计和数据类型；第 2 部分（第 4~6 章）介绍 MySQL 的基本操作，包括 SQL 基础操作、MySQL 连接查询和数据复制、MySQL 基础函数；第 3 部分（第 7、8 章）介绍 MySQL 高级查询函数，包括 MySQL 高级函数和窗口函数；第 4 部分（第 9、10 章）介绍 MySQL 数据表分区，包括 MySQL 表分区和分区管理；第 5 部分（第 11、12 章）介绍视图、存储过程、字符集排序规则的说明和使用，包括 MySQL 视图和存储过程、游标和字符集排序规则；第 6 部分（第 13~15 章）介绍 MySQL 数据查询优化，包括 MySQL 索引、InnoDB 事务模型和锁使用、InnoDB 内存结构解析和优化；第 7 部分（第 16、17 章）介绍 MySQL 数据运维和读写分离架构，包括 MySQL 数据备份和还原、MySQL 主从架构；第 8 部分（第 18 章）介绍 MySQL 数据操作实战。每一部分和章节都是相对独立的主题，读者可以有选择性地单独阅读。此外，本书附有函数速查表，方便读者在实际工作中快速查找函数。

本书适合想要从事大数据开发和提升自身技能的 IT 从业人员、数据库管理员、开发人员等阅读。通过丰富的理论知识和大量的实例，本书将帮助读者更好地应对在实际工作中遇到的与 MySQL 相关的问题。

本书封面贴有清华大学出版社防伪标签，无标签者不得销售。
版权所有，侵权必究。举报：010-62782989，beiqinquan@tup.tsinghua.edu.cn。

图书在版编目（CIP）数据

MySQL 8.0 从入门到实战 / 张文亮编著. 一北京：清华大学出版社，2023.9
ISBN 978-7-302-64561-0

Ⅰ.①M… Ⅱ.①张… Ⅲ.①SQL 语言—数据库管理系统 Ⅳ.①TP311.132.3

中国国家版本馆 CIP 数据核字（2023）第 177839 号

责任编辑：赵 军
封面设计：王 翔
责任校对：闫秀华
责任印制：宋 林

出版发行：清华大学出版社
网　　址：http://www.tup.com.cn，http://www.wqbook.com
地　　址：北京清华大学学研大厦 A 座　　邮　编：100084
社 总 机：010-83470000　　邮　购：010-62786544
投稿与读者服务：010-62776969，c-service@tup.tsinghua.edu.cn
质 量 反 馈：010-62772015，zhiliang@tup.tsinghua.edu.cn

印 装 者：小森印刷霸州有限公司
经　　销：全国新华书店
开　　本：190mm×260mm　　印　张：22.25　　字　数：600 千字
版　　次：2023 年 10 月第 1 版　　印　次：2023 年 10 月第 1 次印刷
定　　价：89.00 元

产品编号：103342-01

前　言

　　MySQL 数据库是目前全球流行的数据库之一。全球最大的网络搜索引擎公司 Google 使用的数据库也是 MySQL，国内的许多大型网络公司，如百度、网易和新浪等，也都选择了 MySQL 数据库。统计数据显示，在全球排名前 20 位的一流互联网公司中，有 80% 都是 MySQL 的忠实用户。目前，MySQL 已被纳入全国计算机等级考试二级的考试科目。

　　在过去十年里，MySQL 社区相继推出了 MySQL 5.6、5.7 和 8.0 版本，这些版本在性能、可扩展性、安全性、稳定性、可维护性和易用性等方面都有了显著的发展。本书以 MySQL 8.0 版本为基础，除了延续之前的经典内容，如架构设计、优化和高可用性等，还增加了 MySQL 过去十年里新版本（包括 5.7 和 8.0 版本）的新特性。

　　本书从实践的角度涵盖数据库系统的架构设计、锁、性能管理、高可用性等内容。除了可作为 MySQL 的参考书籍之外，它也可以作为数据库系统原理和设计的实现参考。本书有助于读者开启数据库的大门，不仅能使读者快速高效地解决业务问题，还能让读者站在数据库管理员的角度来优化自己的数据库。此外，本书还提供了与 MySQL 内部工作原理相关的基本信息和必要解释。读者对 MySQL 的工作原理了解越多，就越能对工程中所涉及的问题进行权衡并做出合理的决定。

　　最后，感谢各位读者选择了本书，希望本书能对读者有所助益。虽然笔者对书中所述内容都尽量进行了核实并进行了多次文字校对，但由于时间紧张和个人水平有限，书中难免存在疏漏之处，敬请广大读者批评指正。笔者会努力采纳读者的意见，争取不断完善本书，以此来回报各位读者对本书的支持。

配套资源下载

本书配套示例源码，需要用微信扫描下边二维码获取。如果阅读中发现问题或疑问，请联系 booksaga@126.com，邮件主题写"MySQL 8.0 从入门到实战"。

笔　者

2023 年 5 月

目 录

第 1 部分 MySQL 基础

第 1 章 初识 MySQL ·· 2
- 1.1 MySQL 是什么 ·· 2
- 1.2 MySQL 8 安装和配置 ·· 3
  - 1.2.1 安装版本的选择 ·· 3
  - 1.2.2 在 Windows 平台下安装和配置 MySQL ························ 3
  - 1.2.3 在 Linux 平台下安装和配置 MySQL ·························· 10
  - 1.2.4 在 Docker 平台下安装和配置 MySQL ························· 14
- 1.3 如何学习 MySQL ·· 18
- 1.4 本章练习 ·· 19

第 2 章 数据库设计 ·· 20
- 2.1 关系数据库设计理论三大范式 ···································· 20
- 2.2 连接数据库和基本信息查询 ······································ 24
  - 2.2.1 连接数据库 ·· 24
  - 2.2.2 基本信息查询 ·· 25
- 2.3 创建数据库和表 ·· 28
  - 2.3.1 创建数据库 ·· 29
  - 2.3.2 创建表 ·· 30
- 2.4 本章练习 ·· 31

第 3 章 MySQL 数据类型 ··· 32
- 3.1 数字数据类型 ·· 32
- 3.2 日期和时间数据类型 ·· 36
- 3.3 字符串数据类型 ·· 37
- 3.4 JSON 数据类型 ··· 43
- 3.5 本章练习 ·· 48

第 2 部分　MySQL 基本操作

第 4 章　MySQL 基础操作 50
- 4.1 数据库操作 50
- 4.2 数据表操作 52
  - 4.2.1 创建数据表 52
  - 4.2.2 数据表的修改与删除 54
  - 4.2.3 约束 55
- 4.3 数据查询 57
- 4.4 数据插入 58
- 4.5 where（筛选） 59
- 4.6 and 和 or（与和或） 60
- 4.7 between（范围查找） 61
- 4.8 distinct（去重关键字） 62
- 4.9 update（修改或更新） 62
- 4.10 delete（删除） 64
- 4.11 like（模糊匹配） 65
- 4.12 order by（排序） 67
- 4.13 group by（分组） 68
- 4.14 with rollup（分组统计再统计） 69
- 4.15 having（分组筛选） 69
- 4.16 union 和 union all（组合查询） 70
- 4.17 本章练习 72

第 5 章　MySQL 连接查询和数据复制 74
- 5.1 连接查询分类 74
- 5.2 inner join（内连接） 75
- 5.3 left join（左连接） 76
- 5.4 right join（右连接） 77
- 5.5 in 和 not in（多关键字筛选） 78
- 5.6 exists 和 not exists（是否返回结果集） 79
- 5.7 复制表 80
- 5.8 临时表 81
- 5.9 本章练习 82

第6章 MySQL 基础函数 ··· 83

6.1 MySQL 字符串函数 ··· 83
6.1.1 character_length(s)：返回字符串长度 ··· 83
6.1.2 concat(s1,s2,…,sn)：字符串合并 ·· 84
6.1.3 format(x,n)：数字格式化 ·· 84
6.1.4 lpad(s1,len,s2)：字符串填充 ··· 84
6.1.5 field(s,s1,s2,…)：返回字符串出现的位置 ································· 84
6.1.6 insert(s1,x,len,s2)：替换字符串 ·· 85
6.1.7 lcase(s)：把字符串中的所有字母转换为小写字母 ···················· 85
6.1.8 ucase(s)：把字符串中的所有字母转换为大写字母 ···················· 85
6.1.9 strcmp(s1,s2)：比较字符串大小 ·· 86
6.1.10 replace(s,s1,s2)：字符串替换 ··· 86
6.1.11 position(s1 in s)：获取子字符串在字符串中出现的位置 ········· 86
6.1.12 md5(s)：字符串加密 ·· 86
6.1.13 inet_aton(ip)：把 IP 地址转换为数字 ····································· 87
6.1.14 inet_ntoa (s)：把数字转换为 IP 地址 ····································· 87

6.2 MySQL 数字函数 ·· 87
6.2.1 ceil(x)：返回不小于 x 的最小整数 ·· 87
6.2.2 ceiling(x)：返回不小于 x 的最小整数 ······································ 88
6.2.3 floor(x)：返回不大于 x 的最大整数 ··· 88
6.2.4 round(x)：返回最接近 x 的整数 ·· 88
6.2.5 max(expression)：求最大值 ··· 89
6.2.6 min(expression)：求最小值 ·· 89
6.2.7 sum(expression)：求总和 ··· 89
6.2.8 avg(expression)：求平均值 ·· 90
6.2.9 count(expression)：求总记录数 ··· 90

6.3 MySQL 日期函数 ·· 91
6.3.1 adddate(d,n)：返回指定日期加上指定天数后的日期 ··············· 91
6.3.2 addtime(t,n)：返回指定时间加上指定时间后的时间 ··············· 91
6.3.3 curdate()：返回当前日期 ·· 92
6.3.4 datediff(d1,d2)：返回两个日期相隔的天数 ······························ 92
6.3.5 dayofyear(d)：返回指定日期是本年的第几天 ·························· 92
6.3.6 extract(type from d)：根据对应格式返回日期 ························· 92
6.3.7 now()：返回当前日期和时间 ·· 93
6.3.8 quarter(d)：返回日期对应的季度数 ··· 94

6.3.9 second(t)：返回指定时间中的秒数 ·· 94
6.3.10 timediff(time1, time2)：计算时间差 ································· 94
6.3.11 date(t)：从指定日期时间中提取日期值 ································ 95
6.3.12 hour(t)：返回指定时间中的小时数 ·································· 95
6.3.13 time(expression)：提取日期时间参数中的时间部分 ······················ 95
6.3.14 time_format(t,f)：根据表达式显示时间 ······························· 95
6.3.15 year(d)：返回指定日期的年份 ····································· 96

第 3 部分　MySQL 高级查询函数

第 7 章　MySQL 高级函数 ·· 98
7.1 cast(x as type)：数据类型转换 ·· 98
7.2 coalesce(expr1, …, exprn)：返回第一个非空表达式 ························· 99
7.3 if(expr,v1,v2)：表达式判断 ··· 99
7.4 ifnull(v1,v2)：null 替换 ·· 99
7.5 isnull(expression)：判断表达式是否为 null ································· 100
7.6 nullif(expr1, expr2)：字符串相等则返回 null ······························ 100
7.7 last_insert_id()：返回最近生成的自增 ID ································· 101
7.8 case expression：表达式分支 ··· 101
7.9 本章练习 ·· 102

第 8 章　MySQL 窗口函数 ·· 103
8.1 窗口函数语法和数据准备 ··· 103
8.2 排名函数 ·· 105
　　8.2.1 row_number()：顺序排名函数 ···································· 105
　　8.2.2 rank()：跳级排名函数 ·· 106
　　8.2.3 dense_rank()：不跳级排名函数 ·································· 107
8.3 分析函数 ·· 108
　　8.3.1 lag()：前分析函数，返回前 n 行的值 ······························ 108
　　8.3.2 lead()：后分析函数，返回后 n 行的值 ······························ 109
8.4 头尾函数 ·· 109
　　8.4.1 first_value()：头尾函数，返回第一个值 ···························· 109
　　8.4.2 last_value()：头尾函数，返回最后一个值 ·························· 110
8.5 聚合统计函数 ·· 112
　　8.5.1 sum()：聚合统计总和 ·· 112
　　8.5.2 avg()：聚合统计平均值 ·· 113

8.5.3 max()：聚合统计最大值 ··· 113
8.5.4 min()：聚合统计最小值 ··· 114
8.5.5 count()：聚合统计总条数 ··· 114
8.6 其他函数 ··· 115
8.6.1 nth_value()：从结果集中的第 n 行获取值 ··· 115
8.6.2 ntile()：数据集分桶 ··· 117
8.7 本章练习 ··· 118

第 4 部分　MySQL 数据表分区

第 9 章　MySQL 表分区 ··· 120
9.1 分区介绍 ··· 120
9.2 分区类型 ··· 121
9.3 范围分区 ··· 121
9.4 列表分区 ··· 124
9.5 列分区 ··· 127
9.5.1 范围列分区 ··· 127
9.5.2 列表列分区 ··· 132
9.6 哈希分区 ··· 134
9.7 键分区 ··· 137
9.8 子分区 ··· 138
9.9 MySQL 分区对 null 的处理 ··· 140
9.10 本章练习 ··· 145

第 10 章　MySQL 分区管理 ··· 146
10.1 范围分区和列表分区的管理 ··· 146
10.1.1 添加和删除分区 ··· 146
10.1.2 拆分合并分区 ··· 149
10.2 哈希分区和键分区的管理 ··· 149
10.3 分区管理和维护操作 ··· 150
10.4 分区的限制 ··· 152
10.5 分区键与主键、唯一键的关系 ··· 153
10.6 本章练习 ··· 156

第 5 部分　视图、存储过程、字符集和排序规则的说明和使用

第 11 章　MySQL 视图和存储过程 ·· 158
11.1　视图 ·· 158
11.2　存储过程 ·· 161
11.2.1　存储过程的创建和调用 ·· 162
11.2.2　带参数的存储过程 ·· 163
11.2.3　变量 ·· 166
11.2.4　注释 ·· 167
11.2.5　存储过程的修改和删除 ·· 168
11.2.6　存储过程的控制语句 ·· 168
11.3　本章练习 ·· 171

第 12 章　MySQL 游标、字符集和排序规则 ·· 172
12.1　游标 ·· 172
12.2　字符集和排序规则 ·· 175
12.2.1　MySQL 中的字符集和排序规则 ··· 175
12.2.2　列字符集和排序规则 ·· 177
12.2.3　使用 collate 查询排序规则 ··· 178
12.3　本章练习 ·· 178

第 6 部分　MySQL 数据查询优化

第 13 章　MySQL 索引 ·· 180
13.1　索引 ·· 180
13.2　创建索引的指导原则 ·· 183
13.3　B-Tree 索引和哈希索引的比较 ··· 184
13.4　SQL 执行计划分析 ··· 185
13.5　MySQL 的索引检索原理 ··· 200
13.5.1　主键索引 ·· 200
13.5.2　普通索引 ·· 204
13.5.3　复合索引 ·· 206
13.6　索引的使用技巧与 SQL 优化实战 ··· 208
13.7　本章练习 ·· 213

第 14 章 InnoDB 存储引擎事务模型和锁的使用 ············· 214

14.1 InnoDB 存储引擎实践 ················ 216
14.2 InnoDB 和 ACID 模型 ················ 217
14.2.1 MySQL 事务处理的两种方法 ········· 218
14.2.2 InnoDB MVCC ··················· 219
14.3 锁机制 ··························· 220
14.3.1 表级锁 ······················· 221
14.3.2 行级锁 ······················· 221
14.3.3 共享锁 ······················· 222
14.3.4 排他锁 ······················· 223
14.3.5 意向锁 ······················· 224
14.3.6 乐观锁 ······················· 225
14.3.7 悲观锁 ······················· 226
14.3.8 间隙锁 ······················· 226
14.3.9 记录锁 ······················· 228
14.3.10 临键锁 ······················ 228
14.3.11 死锁 ······················· 229
14.4 事务隔离级别 ······················ 229
14.5 本章练习 ························· 232

第 15 章 InnoDB 内存结构解析和优化 ·················· 233

15.1 InnoDB 内存结构解析和优化实战 ········· 233
15.1.1 缓冲池原理和优化 ··············· 234
15.1.2 变更缓冲区原理与优化 ············ 245
15.1.3 自适应哈希索引原理与配置 ········· 247
15.1.4 日志缓冲区 ···················· 248
15.2 InnoDB 表实战调优 ·················· 248
15.2.1 优化 InnoDB 表的存储 ············· 248
15.2.2 InnoDB 事务处理优化 ············· 250
15.2.3 InnoDB 只读事务优化 ············· 250
15.2.4 优化 InnoDB 重做日志 ············· 251
15.2.5 InnoDB 磁盘 I/O 优化 ············· 255
15.2.6 配置 InnoDB 线程并发数量 ········· 256
15.2.7 配置 InnoDB 的 I/O 读写后台线程数量 ··· 257
15.2.8 InnoDB 的并行读线程数量 ········· 257

15.2.9　InnoDB 的 I/O 容量·················258

15.3　本章练习·················258

第 7 部分　MySQL 数据运维和读写分离架构

第 16 章　MySQL 数据的备份与还原·················260

16.1　基准测试·················260

16.1.1　Sysbench 介绍·················261

16.1.2　Sysbench 测试实操·················262

16.2　备份·················266

16.3　还原·················267

16.3.1　利用 source 命令导入数据库·················267

16.3.2　利用 load data infile 导入数据·················268

16.3.3　利用 mysqlimport 导入数据·················268

16.4　本章练习·················268

第 17 章　MySQL 主从架构·················269

17.1　读写分离·················269

17.2　MySQL 主从复制·················270

17.3　MySQL 主从复制架构的搭建·················271

17.3.1　配置主节点·················271

17.3.2　配置从节点·················272

17.3.3　主从复制·················273

17.4　本章练习·················274

第 8 部分　MySQL 操作实战

第 18 章　MySQL 数据操作实战·················276

18.1　数据库和数据表的创建与查看·················276

18.2　新增、修改、删除表数据·················280

18.3　基础查询和 where 子句查询·················281

18.4　like 模糊查询·················286

18.5　分组查询、聚合函数、排序查询·················287

18.6　关联查询·················291

18.7　子查询、多表查询·················293

18.8　本章练习·················296

附录 A 函数速查表 ··· 299

A.1 MySQL 字符串函数 ·· 299

- A.1.1 ASCII(s)：返回字符串的 ASCII 码 ······································ 299
- A.1.2 char_length(s)：返回字符串长度 ·· 300
- A.1.3 character_length(s)：返回字符串长度 ································ 300
- A.1.4 concat(s1,s2,…,sn)：字符串合并 ·· 300
- A.1.5 concat_ws(x, s1,s2,…,sn)：字符串合并 ······························ 300
- A.1.6 field(s,s1,s2,…)：返回字符串出现的位置 ····························· 301
- A.1.7 find_in_set(s1,s2)：返回字符串的匹配位置 ························· 301
- A.1.8 format(x,n)：数字格式化 ·· 301
- A.1.9 insert(s1,x,len,s2)：替换字符串 ··· 302
- A.1.10 locate(s1,s)：获取字符串出现的位置 ································ 302
- A.1.11 lcase(s)：把字符串中的所有字母转换为小写字母 ·············· 302
- A.1.12 left(s,n)：返回字符串的前 n 个字符 ································· 302
- A.1.13 lower(s)：把字符串中的所有字母转换为小写字母 ············· 303
- A.1.14 lpad(s1,len,s2)：字符串填充 ·· 303
- A.1.15 ltrim(s)：去掉字符串开始处的空格 ·································· 303
- A.1.16 mid(s,n,len)：从字符串的指定位置截取子字符串 ··············· 303
- A.1.17 replace(s,s1,s2)：字符串替换 ··· 304
- A.1.18 reverse(s)：字符串反转 ··· 304
- A.1.19 right(s,n)：返回字符串后 n 个字符 ··································· 304
- A.1.20 rpad(s1,len,s2)：在字符串中填充相应字符串 ····················· 305
- A.1.21 position(s1 in s)：获取子字符串在字符串中出现的位置 ······ 305
- A.1.22 repeat(s,n)：字符串重复拼接 ··· 305
- A.1.23 substring_index(s, s2, number)：返回字符串中第 n 个出现的子字符串 ······ 305
- A.1.24 trim(s)：去掉字符串开始和结尾处的空格 ························· 306
- A.1.25 ucase(s)：把字符串中的所有字母转换为大写字母 ············· 306
- A.1.26 rtrim(s)：去掉字符串结尾处的空格 ·································· 306
- A.1.27 strcmp(s1,s2)：比较字符串的大小 ··································· 307
- A.1.28 substr(s, start, length)：从字符串中截取子字符串 ·············· 307
- A.1.29 substring(s, start, length)：从字符串中截取子字符串 ········· 307
- A.1.30 md5(s)：字符串加密 ··· 308
- A.1.31 inet_aton (ip)：把 IP 地址转换为数字 ······························ 308
- A.1.32 inet_ntoa(s)：把数字转换为 IP 地址 ································ 308

A.2 MySQL 数字函数 · 309
A.2.1 abs(x)：求绝对值 · 309
A.2.2 sign(x)：返回数字符号 · 309
A.2.3 acos(x)：求反余弦值 · 309
A.2.4 asin(x)：求反正弦值 · 310
A.2.5 atan(x)：求反正切值 · 310
A.2.6 sin(x)：求正弦值 · 310
A.2.7 cos(x)：求余弦值 · 310
A.2.8 cot(x)：求余切值 · 311
A.2.9 tan(x)：求正切值 · 311
A.2.10 degrees(x)：将弧度转换为角度 · 311
A.2.11 radians(x)：将角度转换为弧度 · 312
A.2.12 exp(x)：返回 e 的 x 次方 · 312
A.2.13 ceil(x)：返回不小于 x 的最小整数 · 312
A.2.14 ceiling(x)：返回不小于 x 的最小整数 · 312
A.2.15 floor(x)：返回不大于 x 的最大整数 · 313
A.2.16 round(x)：返回最接近 x 的整数 · 313
A.2.17 greatest(expr1,expr2, …)：返回列表中的最大值 · 313
A.2.18 least(expr1, expr2, expr3, …)：返回列表中的最小值 · 314
A.2.19 ln(x)：求自然对数 · 314
A.2.20 log(x)或 log(base, x)：求对数 · 314
A.2.21 log10(x)：求以 10 为底的对数 · 315
A.2.22 log2(x)：求以 2 为底的对数 · 315
A.2.23 max(expression)：求最大值 · 315
A.2.24 min(expression)：求最小值 · 316
A.2.25 sum(expression)：求总和 · 316
A.2.26 avg(expression)：求平均值 · 317
A.2.27 count(expression)：求总记录数 · 317
A.2.28 mod(x,y)：求余数 · 317
A.2.29 pow(x,y)：求 x 的 y 次方 · 318
A.2.30 sqrt(x)：求平方根 · 318
A.2.31 rand()：求随机数 · 318
A.2.32 truncate(x,y)：返回保留到小数点后 n 位的值 · 319
A.3 MySQL 日期函数 · 319
A.3.1 adddate(d,n)：返回指定日期加上指定天数后的日期 · 319

A.3.2　addtime(t,n)：返回指定时间加上指定时间后的时间 ····································· 319
A.3.3　curdate()：返回当前日期 ··· 320
A.3.4　current_date()：返回当前日期 ·· 320
A.3.5　current_time()：返回当前时间 ·· 320
A.3.6　current_timestamp()：返回当前日期和时间 ·· 320
A.3.7　curtime()：返回当前时间 ·· 321
A.3.8　date(t)：从指定日期时间中提取日期值 ·· 321
A.3.9　datediff(d1,d2)：返回两个日期相隔的天数 ··· 321
A.3.10　date_add(d, interval expr type)：给指定日期加上一个时间段后的日期 ·········· 322
A.3.11　date_format(d,f)：根据表达式显示日期 ··· 322
A.3.12　date_sub(date, interval expr type)：从当前日期减去指定的时间间隔 ············ 322
A.3.13　day(d)：返回日期值中的日值 ·· 323
A.3.14　dayname(d)：返回指定日期是星期几 ·· 323
A.3.15　dayofmonth(d)：返回指定日期是本月的第几天 ······································· 323
A.3.16　dayofweek(d)：返回指定日期是所在星期的第几天 ··································· 323
A.3.17　dayofyear(d)：返回指定日期是本年的第几天 ·· 324
A.3.18　extract (type from d)：根据对应格式返回日期 ·· 324
A.3.19　from_days(n)：返回元年加 n 天的日期 ·· 325
A.3.20　hour(t)：返回指定时间中的小时数 ·· 325
A.3.21　last_day(d)：返回给定日期所在月份的最后一天 ······································ 326
A.3.22　localtime()：返回当前日期和时间 ··· 326
A.3.23　localtimestamp()：返回当前日期和时间 ··· 326
A.3.24　makedate(year, day-of-year)：时间组合天数 ·· 326
A.3.25　maketime(hour, minute, second)：时间组合 ··· 327
A.3.26　microsecond(date)：返回时间参数中的微秒数 ··· 327
A.3.27　minute(t)：返回指定时间中的分钟数 ··· 327
A.3.28　monthname(d)：返回日期中的月份名称 ··· 328
A.3.29　month(d)：返回日期中的月份值 ·· 328
A.3.30　now()：返回当前日期和时间 ··· 328
A.3.31　period_add(period, number)：日期加月份数 ·· 328
A.3.32　period_diff(period1, period2)：返回两个时间之间的月份差 ·························· 329
A.3.33　quarter(d)：返回日期对应的季度数 ·· 329
A.3.34　second(t)：返回指定时间中的秒数 ··· 329
A.3.35　sec_to_time(s)：秒数转换为时间 ·· 329
A.3.36　str_to_date(string, format_mask)：把字符串转换为日期 ······························ 330

A.3.37 subdate(d,n): 从指定日期减去指定天数后的日期 ································ 330
A.3.38 subtime(t,n): 从指定时间减去指定秒数后的时间 ································ 330
A.3.39 sysdate(): 返回当前日期和时间 ·· 331
A.3.40 time(expression): 提取日期时间参数中的时间部分 ································ 331
A.3.41 time_format(t,f): 根据表达式显示时间 ··· 331
A.3.42 time_to_sec(t): 把时间转换为秒数 ·· 331
A.3.43 timediff(time1, time2): 计算时间差 ··· 332
A.3.44 timestamp(expression, interval): 指定时间加上时间间隔后的时间 ······· 332
A.3.45 to_days(d): 计算元年到当前日期的天数 ··· 332
A.3.46 week(d): 返回指定日期是本年的第几周 ··· 333
A.3.47 weekday(d): 返回指定日期是星期几 ··· 333
A.3.48 year(d): 返回指定日期的年份 ·· 333

A.4 MySQL 高级函数 ·· 333

A.4.1 bin(x): 返回字符串的二进制编码 ··· 333
A.4.2 binary(s): 将字符串转换为二进制数 ·· 334
A.4.3 case expression: 表达式分支 ··· 334
A.4.4 cast(x as type): 数据类型转换 ·· 335
A.4.5 coalesce(expr1,…,exprn): 返回第一个非空表达式 ······································· 335
A.4.6 connection_id(): 返回唯一连接 ID ·· 335
A.4.7 conv(x,f1,f2): 数据进制转换 ··· 335
A.4.8 convert(s using cs): 求字符串的字符集 ·· 336
A.4.9 current_user(): 返回当前的用户名 ·· 336
A.4.10 database(): 返回当前的数据库名 ·· 336
A.4.11 if(expr,v1,v2): 表达式判断 ·· 337
A.4.12 ifnull(v1,v2): null 替换 ··· 337
A.4.13 isnull(expression): 判断表达式是否为 null ·· 337
A.4.14 nullif(expr1, expr2): 字符串相等则返回 null ··· 338
A.4.15 last_insert_id(): 返回最近生成的自增 ID ··· 338
A.4.16 session_user(): 返回当前会话的用户名 ·· 339
A.4.17 version(): 返回数据库的版本号 ··· 339

第 1 部分　MySQL 基础

第 1 章　初识 MySQL
第 2 章　数据库设计
第 3 章　MySQL 数据类型

第 1 章

初识 MySQL

本章内容：
- MySQL 的介绍
- 在 Windows 上安装 MySQL
- 在 Linux 上安装 MySQL
- 在 Docker 上安装 MySQL

本章将讲解 MySQL 在 Web 业务场景下的意义和优势，介绍 MySQL 8 的特性和作用，并讲述在多种环境下安装 MySQL 的步骤，以便初学者可以快速开始 MySQL 的学习。

1.1 MySQL 是什么

MySQL 是当下流行的关系数据库管理系统（Relational Database Management System，RDBMS），使用 C 和 C++语言编写而成，因而具有源码级的可移植性。MySQL 可以在 Linux、macOS、Novell NetWare、OpenBSD、Solaris、Windows 等多种操作系统中运行。在 Web 应用方面，MySQL 是非常好的关系数据库管理系统应用软件之一。作为一种关系数据库管理系统，它将数据保存在不同的数据表中。MySQL 所使用的 SQL 语言是用于访问数据库的常用标准化语言。由于 MySQL 体积小、速度快、总拥有成本低，尤其是具有开放源码这一特点，一般中小型网站的开发都选择 MySQL 作为网站数据库。由于其社区版性能卓越，因此搭配各种后端语言可以组成良好的开发环境。此外，MySQL 既可以嵌入应用程序中，也可以独立支持数据仓库、全文索引和高可用的冗余系统、在线事务处理系统。另外，MySQL 还具有如下特点：

- 为多种编程语言提供了 API，这些编程语言包括 C、C++、Python、Java、Perl、PHP、Ruby、.NET 等。
- 支持多线程，可以充分利用 CPU 资源。
- 具有优化的 SQL 查询算法，有效地提高了查询速度。
- 提供 TCP/IP、ODBC 和 JDBC 等多种数据库连接途径。
- 支持可以处理上千万条记录的大型数据库。
- 支持多种存储引擎，比如 InnoDB、MyISAM 等。
- MySQL 是可以定制的，它采用了 GPL 协议，因而用户可以通过修改源码来开发自己的 MySQL 系统。
- 提供多语言支持，常见的编码如中文的 GB2312、UTF-8 等都可以用作数据表名和数据列名。
- 提供用于管理、检查、优化数据库操作的管理工具。
- 具有在线 DDL 更改功能，数据架构支持动态应用程序，提高了开发人员操作数据表的灵活性。
- 复制无崩溃从机，可以提高可用性。
- 复制多线程从机，可以提高性能。

1.2　MySQL 8 安装和配置

1.2.1　安装版本的选择

在安装 MySQL 时，应选择自己所需的版本和相应的文件格式。MySQL 有开发版本和通用版本。开发版本具有最新功能，但不建议用于生产用途。通用版本可用于生产用途，因此建议安装最新的版本。

MySQL 8.0 的版本号由三个数字和一个可选后缀组成，例如 MySQL-8.0.3.[x]，x 可选，如果有，则会随着版本的发布而自增。版本号中的数字解释如下：

- 第一个数字（8）是主版本号。
- 第二个数字（0）是次版本号。主、次版本号一起构成发布的序列号，序列号表示包含稳定的功能集。
- 第三个数字（3）是发布系列中的版本号。每个新的错误修订版的版本号中都会增加此值。一般情况下建议安装最新的版本。

1.2.2　在 Windows 平台下安装和配置 MySQL

MySQL 只能安装在 Microsoft Windows 64 位操作系统中，如果想要安装 MySQL 8.0 Server（服务器版），那么还需要在系统中安装 Microsoft Visual C++ 2015 Redistributable Package 和 Microsoft .NET Framework 4.5.2 或更高版本。在安装服务器版本之前，应确保在本地系统中已

下载好软件安装包。下面将演示在 Windows 系统中安装 MySQL 数据库的两种方式。

1. MySQL Installer（利用安装程序安装）

步骤01 进入 MySQL 官网下载相关安装包，如图 1-1 所示。第一个安装包用于在线安装，在网络畅通的情况下，可以选用这种安装方式。第二个是可以离线安装的软件包，一般建议采用这种安装方式。

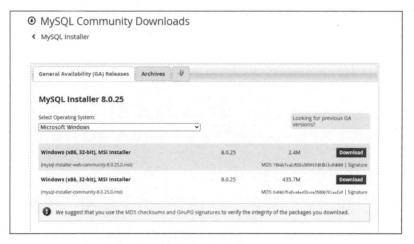

图 1-1　下载安装包

步骤02 首次下载 MySQL 安装程序时，安装向导会引导我们完成 MySQL 产品的初始安装。首次安装需要进行初始设置，MySQL 安装程序在初始设置期间会检测主机上安装的现有 MySQL 产品，并将它们添加到要管理的产品列表中。如图 1-2 所示，当我们打开 MSI 文件之后，会提示我们选择安装的类型。

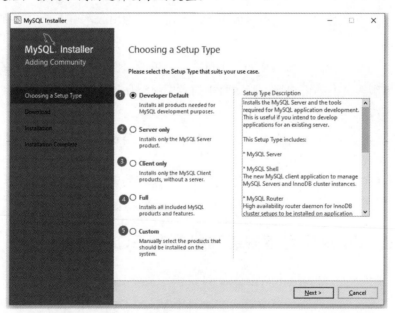

图 1-2　选择安装的类型

如图 1-2 所示，总共有 5 种安装类型。

① Developer Default（默认安装）：用于 MySQL 应用程序的开发。如果我们想要利用 MySQL 来进行应用程序的开发，可以选择此安装类型。

② Server only（仅安装服务器）：仅安装 MySQL 服务器。此安装类型在下载 MySQL 服务器安装包时会让用户选择安装 GA 版（一般可用性，即稳定版）或开发版。该安装类型使用默认的安装路径和数据存储路径。

③ Client only（仅安装客户端）：仅安装最新的 MySQL 应用程序和 MySQL 连接器。此安装类型类似于默认安装类型，不同之处在于它不包括 MySQL 服务器或通常与服务器捆绑的客户端程序，例如 mysql 或 mysqladmin。

④ Full（完整）：安装所有可用的 MySQL 产品。如果是初学者，那么可以选择此安装类型。

⑤ Custom（自定义）：此安装类型可以从 MySQL 安装程序目录中筛选想安装的各个 MySQL 产品。

步骤 03 选择好想要的安装类型，单击 Next 按钮，就会进入安装需求检查对话框，如图 1-3 所示。

图 1-3 安装需求检查对话框

在安装需求检查对话框中包括如下内容：

① 显示初始设置中的当前步骤。读者在此列表中看到的步骤可能会略有不同，具体步骤取决于主机上已安装的产品、必备软件的可用性以及读者想要安装在主机上的产品。

② 按产品列出所有待确认的安装需求。

③ 安装需求的详细说明可以帮助我们解决安装前需要解决的问题。如果需要安装必备的软件，则该界面会提供下载用的网址（URL）。在下载并安装所需的软件后，单击 Check 按钮来验证是否已经满足安装需求。

④ 提供以下辅助安装的操作：

- Back：返回上一步。此操作可以让我们回到上一步重新选择安装类型。
- Execute：让 MySQL 安装程序尝试为所有选择的安装项目下载和安装必备的软件。
- Next：不进行安装需求检查而继续安装产品，其中不包括未通过安装需求检查的产品。
- Cancel：取消安装 MySQL 产品。

步骤 04 MySQL 网络和身份认证配置，如图 1-4 所示。

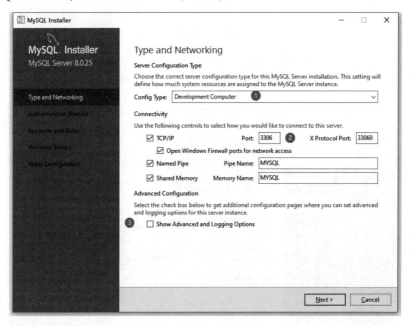

图 1-4　MySQL 网络配置

在该对话框中有以下选项：

① 选择需要配置的服务类型。

② 默认的服务端口号，①和②这两个配置可以不做修改。

③ 建议勾选 Show Advanced and Logging Options 复选框，以便在后续步骤中设置自定义日志记录和高级选项。

继续往下执行，会出现设置密码的对话框，如图 1-5 所示。

图 1-5　账户密码设置

该对话框包含以下内容：

① 给服务设置密码。

② 重复输入一次以确认密码，应当确保这两次输入的密码一致。

步骤 05　MySQL 服务名称设置，如图 1-6 所示。

图 1-6　MySQL 服务名称设置

在该对话框中输入 MySQL 服务的名称，建议使用默认值。

步骤06 设置 MySQL 服务相关的输出日志存储的路径，如图 1-7 所示。

图 1-7　设置 MySQL 服务输出日志存储的路径

在该对话框中包含以下内容：

① 设置 MySQL 服务的错误日志输出路径。

② 设置一般的日志输出路径。

③ 设置查询日志的输出路径。

④ 设置二进制日志的输出路径。

步骤07 验证是否安装成功。如图 1-8 所示，按住键盘上的 Windows 键，选择"运行"，然后在弹出的对话框中输入 services.msc，并按回车键。

接下来会出现如图 1-9 所示的服务窗口。如果可以看到 MySQL 8.0 这个服务，就表示 MySQL 数据库安装成功。

图 1-8　进入"运行"对话框

图 1-9　系统服务窗口

在服务成功启动之后，可以使用界面化工具 Navicat 连接数据库，当出现如图 1-10 所示的对话框时，输入主机名、端口号以及安装时设置的密码。在单击"测试连接"按钮之后，若出现连接成功的字样，则说明安装与配置都正确。

图 1-10　通过界面化工具测试连接

2. 在 Windows 上默认安装 MySQL 服务

表 1-1 所示是以默认安装类型安装 MySQL 服务后生成的相关目录，当我们需要查看相关信息或者数据日志时可能会用到。

表 1-1 默认安装 MySQL 服务之后生成的相关目录

目　　录	目录内容	备　　注
bin	mysqld 服务器、客户端和应用程序	
%PROGRAMDATA%\MySQL\MySQL Server 8.0\	日志文件、数据库文件	Windows 系统变量%PROGRAMDATA% 默认为 C:\ProgramData
docs	发布文档	使用 MySQL 安装程序，使用 Modify 操作可选择此文件夹
lib	服务相关的程序集	
share	字符集文件示例配置文件，用于数据库安装的 SQL	

1.2.3　在 Linux 平台下安装和配置 MySQL

使用 MySQL Yum 存储库安装最新的 MySQL GA 版本的步骤如下。

1. 添加 MySQL Yum 存储库

要将 MySQL Yum 存储库添加到系统的存储库列表中，可以使用 RPM 命令来完成，具体步骤如下：

首先选择并下载适用于目标平台的安装包（发布包）。

我们可以查看目标系统的平台版本（操作系统版本序列号）。如图 1-11 所示，该平台的版本序列号为 7。

```
[root@localhost home]# cat /etc/redhat-release
CentOS Linux release 7.5.1804 (Core)
[root@localhost home]#
```

图 1-11　查看目标系统的操作系统版本序列号

然后下载工具 wget，下载完成之后用下面的命令进行安装：

```
yum install wget -y
```

接着下载对应版本的 RPM 文件，因为目前的平台是 7 系列，所以选择安装 7.3 版本的 RPM 文件。执行如下命令进行下载：

```
wget https://dev.mysql.com/get/mysql80-community-release-el7-3.noarch.rpm
```

当出现如图 1-12 所示的界面时，说明下载成功。

```
[root@localhost home]# wget https://dev.mysql.com/get/mysql80-community-release-el7-3.noarch.rpm
--2021-06-02 22:09:32--  https://dev.mysql.com/get/mysql80-community-release-el7-3.noarch.rpm
Resolving dev.mysql.com (dev.mysql.com)... 137.254.60.11
Connecting to dev.mysql.com (dev.mysql.com)|137.254.60.11|:443... connected.
HTTP request sent, awaiting response... 302 Found
Location: https://repo.mysql.com/mysql80-community-release-el7-3.noarch.rpm [following]
--2021-06-02 22:09:34--  https://repo.mysql.com/mysql80-community-release-el7-3.noarch.rpm
Resolving repo.mysql.com (repo.mysql.com)... 104.75.165.42
Connecting to repo.mysql.com (repo.mysql.com)|104.75.165.42|:443... connected.
HTTP request sent, awaiting response... 200 OK
Length: 26024 (25K) [application/x-redhat-package-manager]
Saving to: 'mysql80-community-release-el7-3.noarch.rpm.1'

100%[======================================================>] 26,024      --.-K/s   in 0.1s

2021-06-02 22:09:34 (223 KB/s) - 'mysql80-community-release-el7-3.noarch.rpm.1' saved [26024/26024]
```

图 1-12　下载 RPM 包

2. 安装下载的发布包

安装命令如下：

```
yum install platform-and-version-specific-package-name.rpm
```

需要注意的是，platform-and-version-specific-package-name 为下载的 RPM 包的名称。而后执行如下命令：

```
yum install mysql80-community-release-el7-3.noarch.rpm -y
```

如果出现如图 1-13 所示的 Complete 说明，就表示发布包安装完成。

```
[root@worker1 ~]# yum install mysql80-community-release-el7-3.noarch.rpm -y
Loaded plugins: fastestmirror
Examining mysql80-community-release-el7-3.noarch.rpm: mysql80-community-release-el7-3.noarch
Marking mysql80-community-release-el7-3.noarch.rpm to be installed
Resolving Dependencies
--> Running transaction check
---> Package mysql80-community-release.noarch 0:el7-3 will be installed
--> Finished Dependency Resolution

Dependencies Resolved

================================================================================
 Package                        Arch      Version    Repository                                      Size
================================================================================
Installing:
 mysql80-community-release      noarch    el7-3      /mysql80-community-release-el7-3.noarch         31 k

Transaction Summary
================================================================================
Install  1 Package

Total size: 31 k
Installed size: 31 k
Downloading packages:
Running transaction check
Running transaction test
Transaction test succeeded
Running transaction
  Installing : mysql80-community-release-el7-3.noarch                           1/1
  Verifying  : mysql80-community-release-el7-3.noarch                           1/1

Installed:
  mysql80-community-release.noarch 0:el7-3

Complete!
```

图 1-13　发布包安装成功

如果使用的是其他系列的版本，那么可以根据实际情况选择安装相应的发布包。

对于基于 EL6 的系统，命令的形式为：

```
yum install mysql80-community-release-el6-{version-number}.noarch.rpm
```

对于基于 EL7 的系统,命令的形式为:

```
yum install mysql80-community-release-el7-{version-number}.noarch.rpm
```

对于基于 EL8 的系统,命令的形式为:

```
yum install mysql80-community-release-el8-{version-number}.noarch.rpm
```

对于 Fedora 34,命令的形式为:

```
dnf install mysql80-community-release-fc34-{version-number}.noarch.rpm
```

对于 Fedora 33,命令的形式为:

```
dnf install mysql80-community-release-fc33-{version-number}.noarch.rpm
```

3. 安装 MySQL

通过以下命令安装 MySQL:

```
yum install mysql-community-server -y
```

这条命令将安装 MySQL 服务器包(mysql-community-server)以及运行该服务器所需组件的包,包括客户端包(mysql-community-client)、客户端和服务器的常见错误消息和字符集(mysql-community-common)以及共享客户端库(mysql-community-libs)。如果出现如图 1-14 所示的信息,就表示安装成功。

图 1-14　安装 MySQL

4. 启动 MySQL 服务器

使用以下命令启动 MySQL 服务器:

```
systemctl start mysqld
```

使用以下命令检查 MySQL 服务器的状态:

```
systemctl status mysqld
```

查询服务的状态,显示 running 时表示服务已经启动,如图 1-15 所示。

```
[root@localhost ~]# systemctl status mysqld
● mysqld.service - MySQL Server
   Loaded: loaded (/usr/lib/systemd/system/mysqld.service; enabled; vendor preset: disabled)
   Active: active (running) since Mon 2021-05-31 23:51:45 EDT; 8s ago
     Docs: man:mysqld(8)
           http://dev.mysql.com/doc/refman/en/using-systemd.html
  Process: 4502 ExecStartPre=/usr/bin/mysqld_pre_systemd (code=exited, status=0/SUCCESS)
 Main PID: 4631 (mysqld)
   Status: "Server is operational"
    Tasks: 38
   Memory: 470.1M
   CGroup: /system.slice/mysqld.service
           └─4631 /usr/sbin/mysqld

May 31 23:51:36 localhost.localdomain systemd[1]: Starting MySQL Server...
May 31 23:51:45 localhost.localdomain systemd[1]: Started MySQL Server.
[root@localhost ~]#
```

图 1-15　MySQL 服务状态

需要注意的是，如果操作系统已启用 systemd，则应使用标准 systemctl（或 service）命令（例如 stop、start、status 和 restart）来管理 MySQL 服务器的服务。mysqld 服务默认是启用的（在系统重新启动时启用了）。

5. 登录 MySQL

在安装 MySQL 服务时系统创建了一个超级用户账户（root）。超级用户的密码已设置并存储在错误日志文件中，如果我们想要知道密码，可以使用如下命令进行查询：

```
grep 'temporary password' /var/log/mysqld.log
```

如图 1-16 所示，最后输出的字符串就是想要查看的密码。

```
[root@localhost ~]# grep 'temporary password' /var/log/mysqld.log
2021-06-01T03:51:40.448706Z 6 [Note] [MY-010454] [Server] A temporary password is generated for root@localhost: T<kB=Rl_,05f
[root@localhost ~]#
```

图 1-16　想要查看的密码

我们使用生成的临时密码登录并为超级用户账户设置自定义密码，以便尽快更改 root 密码。首先执行下面的命令进行登录：

```
mysql -uroot -p
```

如图 1-17 所示，在登录过程中需要输入密码，可以输入之前查询到的密码进行登录。

```
[root@localhost ~]# mysql -uroot -p
Enter password:
Welcome to the MySQL monitor.  Commands end with ; or \g.
Your MySQL connection id is 8
Server version: 8.0.25

Copyright (c) 2000, 2021, Oracle and/or its affiliates.

Oracle is a registered trademark of Oracle Corporation and/or its
affiliates. Other names may be trademarks of their respective
owners.

Type 'help;' or '\h' for help. Type '\c' to clear the current input statement.

mysql>
```

图 1-17　MySQL 登录

然后执行如下命令修改密码：

```
alter user 'root'@'localhost' identified by 'NewPass';
```

在修改密码时注意密码要尽量复杂，需要包含数字、大小写字母和标记符号，否则系统会出现如图 1-18 所示的提示，提示密码修改出错。

图 1-18　修改 MySQL 账户的密码

1.2.4　在 Docker 平台下安装和配置 MySQL

Docker 是一个用于开发、交付和运行应用程序的开放平台。Docker 能够将应用程序与基础架构分开，从而快速交付软件。借助 Docker，可以采用与管理应用程序相同的方式来管理基础架构。

Docker 是当下热门的容器，为了省去 Linux 下安装配置程序烦琐且易于出错的步骤，笔者增加了本节内容，安装 Docker 环境后，就能在其下快速地安装 Redis 服务。

1. 在 CentOS 下安装 Docker

安装 Docker 的系统需求是必须具备一个 CentOS 7 或者 CentOS 8 的维护版本，不支持之前的旧版本。

1）卸载旧版本

较旧的 Docker 版本称为 docker 或 docker-engine。如果安装过这些程序，请卸载它们及其相关的依赖项。

```
$ sudo yum remove docker \
        docker-client \
        docker-client-latest \
        docker-common \
        docker-latest \
        docker-latest-logrotate \
        docker-logrotate \
        docker-engine
```

2）更新系统 yum 工具包

需要安装 yum-utils 软件包：

```
$ sudo yum install -y yum-utils
```

3）设置稳定的存储库（可选择其中一个）

可以根据自己的网络情况选择存储库地址，推荐使用阿里云和清华大学源，如果设置官方源，那么需要单独设置好网络才可以进行正常操作。

使用官方源地址：

```
$ sudo yum-config-manager \
   --add-repo \
https:// download.docker.com/linux/centos/docker-ce.repo
```

使用阿里云：

```
$ sudo yum-config-manager \
   --add-repo \
http:// mirrors.aliyun.com/docker-ce/linux/centos/docker-ce.repo
```

使用清华大学源：

```
$ sudo yum-config-manager \
   --add-repo \
https:// mirrors.tuna.tsinghua.edu.cn/docker-ce/linux/centos/docker-ce.repo
```

4）安装 Docker Engine-Community

安装社区版 Docker 包：

```
$ sudo yum install docker-ce docker-ce-cli containerd.io
```

5）启动 Docker

启动 Docker 很简单：

```
$ sudo systemctl start docker
```

6）验证是否正确安装

可以使用如下命令来验证 Docker 是否正确安装：

```
$ sudo docker run hello-world
```

2. 在 Windows 10 下安装 Docker

Docker Desktop 是 Docker 在 Windows 10 和 macOS 操作系统上的官方安装方式，这种方式依然是先在虚拟机中安装 Linux，再安装 Docker。用户可以从网上搜索 Docker Desktop 的安装软件 docker-ce-desktop-windows（此方法仅适用于 Windows 10 操作系统专业版、企业版、教育版和部分家庭版）。

1）安装 Hyper-V 虚拟机

Hyper-V 是微软开发的虚拟机，类似于 VMWare 或 VirtualBox，只是 Hyper-V 仅适用于 Windows 10。这是 Docker Desktop for Windows 所使用的虚拟机（这个虚拟机一旦启用，VirtualBox、VMWare、Workstation 15 及以下版本将无法使用。如果必须在计算机上使用其他虚拟机，则不要在 Windows 中启动 Hyper-V）。

2）启动 Hyper-V

启动 Hyper-V 的操作步骤如下：

步骤 01 在 Windows 中右击"开始"按钮，在弹出的快捷菜单中选择"应用和功能"命令，如图 1-19 所示。

步骤 02 打开"程序和功能"窗口，单击"启用或关闭 Windows 功能"链接，如图 1-20 所示。

图 1-19 选择"应用和功能"命令　　　　图 1-20 "程序和功能"窗口

步骤 03 在打开的"Windows 功能"对话框中勾选"Hyper-V"复选框，然后单击"确定"按钮，如图 1-21 所示。

图 1-21 勾选"Hyper-V"复选框

3)安装 Docker

接下来双击下载的安装 Docker 的 EXE 文件,在打开的对话框中单击 Next 按钮,最后单击 Finish 按钮完成安装。

安装完成后,Docker 会自动启动,通知栏上会出现 图标,表示 Docker 正在运行,然后按 Win+R 快捷键,在弹出的"运行"窗口中输入 PowerShell。

用户还可以执行命令 docker run hello-world 来检查是否安装成功。

3. 利用 Docker 安装 MySQL

1)下载 MySQL 镜像

执行 docker pull mysql/mysql-server:8.0 下载 MySQL 镜像,如图 1-22 所示。

图 1-22 下载镜像

2)启动 MySQL 服务

执行如下命令启动 MySQL 服务:

```
docker run --name=mysql1 --restart on-failure -d mysql/mysql-server:8.0
```

3)其他操作

删除容器的命令如下:

```
docker rm -f mysql1
```

停止容器的命令如下:

```
docker stop mysql1
```

4)查看启动的日志

查看启动日志的命令如下:

```
docker logs mysql1
```

查看 MySQL 服务的密码的命令如下:

```
docker logs mysql1 2>&1 | grep GENERATED
```

查询到的密码如图 1-23 所示。

图 1-23 查询到的密码

5）登录 MySQL 服务

登录 MySQL 服务的命令如下：

```
docker exec -it mysql1 mysql -uroot -p
```

当系统询问时，输入生成的 root 密码。由于该 MYSQL_ONETIME_PASSWORD 选项默认为 true，因此在将 MySQL 客户端连接到服务器后，必须通过以下语句来重置服务器 root 账户的密码：

```
alter user 'root'@'localhost' identified by 'password';
```

或者执行下面的命令进行安装：

```
docker run -p 3306:3306 --name mysql \
-v /root/mysql/log:/var/log/mysql \
-v /root/mysql/data:/var/lib/mysql \
-v /root/mysql/conf:/etc/mysql \
-e MYSQL_ROOT_PASSWORD=123456 \
-d mysql/mysql-server:8.0
```

以上命令挂载了服务的日志文件和 lib 文件，已经设置好了密码。需要注意的是，在执行这条命令之前，要在宿主机的/root/mysql/conf 目录下创建 my.cnf 文件，这个文件是服务启动的配置文件，该配置文件至少应该有如下几行内容：

```
[mysqld]
init_connect='set collation_connection = utf8_general_ci'
init_connect='set names utf8'
secure_file_priv=
```

1.3　如何学习 MySQL

在学习 MySQL 数据库之前，很多读者会问，如何才能更好地学习 MySQL 的相关技能呢？下面就讲述学习 MySQL 的方法。

1. 培养兴趣

兴趣是最好的老师，不论学习什么知识，兴趣都可以极大地提高学习效率，学习 MySQL 也不例外。

2. 扎实基础

随着技术应用的不断深入，只有有着扎实的基础功底，才能在技术道路上走得更远。对于 MySQL 的学习而言，SQL 语句是其中最基础的部分，很多功能都通过 SQL 语句来实现，所以读者在学习的过程中要多编写 SQL 语句，最好能对同一功能使用不同的 SQL 语句来实现。

1.4 本章练习

1. 下载并安装 MySQL 数据库。
2. 使用配置向导配置 MySQL 为系统服务，并设置为手动启动或者关闭 MySQL 服务。

第 2 章

数据库设计

本章内容:
- 数据库设计理论
- 连接数据库
- 创建数据库和表

本章将讲解数据库设计理论,以简单的示例讲解三大范式以及 MySQL 中对数据库和表的基本操作(包括创建数据库、切换数据库、创建表等)。

2.1 关系数据库设计理论三大范式

关系模型(Relational Model)是一种基于表的数据模型,它广泛应用于数据库管理系统(DBMS)中。关系模型的核心思想是将数据组织成表,每个表包含一组相关的行和列。这些行和列之间的关系通过主键和外键来定义,从而实现数据的完整性、一致性和可维护性。关系模型中一些重要术语介绍如下:

- 属性(Attribute):列的名称。
- 关系(Relation):列属性之间存在的某种关联。
- 表(Table):由多个属性以及众多元组所表示的各个实例组成。
- 键(Key):由一个或多个属性组成,其值能唯一标识关系中的一个元组。如果某个关系 A 中的一个(组)属性是另一个关系 B 的键,则该(组)属性在 A 中被称为外键。
- 笛卡儿积(交叉连接,Cross Join):第一个关系的每一行数据分别与第二个关系的每一行数据组合。
- 自然连接(Natural Join):第一个关系的每一行数据与第二个关系的每一行数据进行

匹配，若得到交叉部分则合并，若无交叉部分则舍弃。
- θ 连接（Theta Join）：加上约束条件的笛卡儿积，先得到笛卡儿积，再根据约束条件删除不满足条件的元组。
- 外连接（Outer Join）：执行自然连接后，将舍弃的部分也加入，并且把匹配失败处的属性设置为 null。

表 2-1 是一张酒店客人入住信息表。

表2-1　酒店客人入住信息表

客人编号	姓名	地址	……	客房号	客房描述	客房类型	客房状态	床位数	价格（元）	入住人数
C1001	张三	Addr1	……	1001	A 栋 1 层	单人间	入住	1	128.00	1
C1002	李四	Addr2	……	2002	B 栋 2 层	标准间	入住	2	168.00	0
C1003	王五	Addr3	……	2002	B 栋 2 层	标准间	入住	2	168.00	2
C1004	赵六	Addr4	……	2003	B 栋 2 层	标准间	入住	2	158.00	1
……	……	……	……	……	……	……	……	……	……	……
C8006	A1	Addrm	……	101	C 栋 3 层	总统套房	入住	3	1080.00	1
C8008	A2	Addrn	……	101	C 栋 3 层	总统套房	空闲	3	1080.00	0

通过表 2-1 可以发现，该表存在不符合规范的设计，总共有如下几点：

（1）信息重复：比如客房类型和客房状态存在大量的数据重复。

（2）更新异常：修改了一个记录中的信息，但是另一个记录中相同的信息却没有被同步修改。

（3）插入异常：无法正确表示信息。

（4）删除异常：丢失有效信息。

下面利用三大范式对数据库进行改造。

第一范式（1NF）：目标是确保每列的原子性，如果每列都是不可再分的最小数据单元（也被称为最小的原子单元），则满足第一范式。

例 2.1　第一范式优化表格示例。

没有经过第一范式优化的表格如表 2-2 所示。

表2-2　没有经过第一范式优化的表格

Address
中国北京
美国纽约
英国利物浦
日本东京
……

经过第一范式优化的表格如表 2-3 所示。

表 2-3 经过第一范式优化的表格

CustID	Country	City
1	中国	北京
3	英国	利物浦
4	日本	东京
2	美国	纽约
……	……	……

第二范式（2NF）：要求每张表只描述一件事情。

例 2.2 第二范式优化表格示例。

没有经过第二范式优化的表格如表 2-4 所示。

表 2-4 没有经过第二范式优化的表格

字 段	例 子
客人编号	C1002
姓名	李四
地址	Addr1
客房号	2002
客房状态	入住
客房类型	标准间
床位数	2
入住人数	2
价 格	168
……	……

经过第二范式优化之后，把一张表拆解成两张表，分别为 Guest 表和 Room 表。

Guest 表如表 2-5 所示。

表 2-5 Guest 表

字 段	例 子
客人编号	C1002
姓名	李四
地址	Addr1
客房号	2002

Room 表如表 2-6 所示。

表 2-6 Room 表

字 段	例 子
客房号	2002
客房状态	入住
客房类型	标准间
床位数	2
入住人数	2

第三范式（3NF）：如果一个关系满足第二范式，并且除了主键以外的其他列都不依赖于主键列，则满足第三范式。

例 2.3 第三范式优化表格示例。

没有经过第三范式优化的 Room 表如表 2-7 所示。

表2-7　没有经过第三范式优化的Room表

字　段	例　子
客房号	2002
客房状态	入住
客房类型	标准间
床位数	2
入住人数	2
价格	168
……	……

经过第三范式优化之后，拆成 3 张表，分别是 Room 表、RoomType 表和 RoomState 表。Room 表如表 2-8 所示。

表2-8　Room表

字　段	例　子
客房号	2002
客房状态编号	1
客房类型编号	2
入住人数	2
……	……

RoomType 表如表 2-9 所示。

表2-9　RoomType表

字　段	例　子
客房类型编号	2
客房类型名称	标准间
床位数	2
价格	168

RoomState 表如表 2-10 所示。

表2-10　RoomState表

字　段	例　子
客房状态编号	1
客房状态名称	入住

如上所述，经过三大范式将一张存在问题的大表拆分成 4 张表，最终数据库设计结构如图 2-1 所示（PK 表示主键，FK 表示外键）。

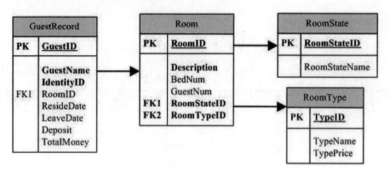

图 2-1　酒店管理系统数据库模型

2.2　连接数据库和基本信息查询

数据库设计好了，本节就来连接数据库并进行基本信息查询。

2.2.1　连接数据库

要连接到 SQL 服务器，需要在调用 MySQL 时提供用户名，并且很可能需要提供密码。如果 SQL 服务器在其他计算机上运行，还必须指定主机名。此时应联系管理员，了解该使用哪些连接参数来进行连接（即使用的主机、用户名和密码）。当知道正确的参数之后，可以执行如下命令进行连接：

```
shell> mysql -h host -u user -p
Enter password: ********
```

host 和 user 分别代表运行的 MySQL 服务器的主机名和 MySQL 账户的用户名。******** 代表密码，当 MySQL 显示"Enter password:"提示时，输入账户对应的密码。如果输入密码正确，那么会看到如下信息：

```
shell> mysql -h host -u user -p
Enter password: ********
Welcome to the MySQL monitor.  Commands end with ; or \g.
Your MySQL connection id is 25338 to server version: 8.0.25-standard
Type 'help;' or '\h' for help. Type '\c' to clear the buffer.
mysql>
```

出现"mysql>"提示符就表示可以输入 SQL 语句了。
如果在运行 MySQL 的同一台计算机上登录，则可以省略主机名，只需执行以下命令：

```
shell> mysql -u user -p
```

如果在尝试登录时收到错误提示信息，例如 ERROR 2002 (HY000): Can't connect to local MySQL server through socket '/tmp/mysql.sock' (2)，则表示 MySQL 服务器守护程序（在 UNIX 操作系统中）或服务（在 Windows 操作系统中）未运行，也就是需要启动 MySQL 服务。

2.2.2 基本信息查询

本小节将介绍信息查询的基本原则，通过几个查询示例来熟悉 MySQL 的工作原理。

例 2.4　查询服务器的版本号和当前日期。在"mysql>"提示符后面输入如下命令，然后按 Enter 键：

```
mysql> select version(), current_date;
+-----------+--------------+
| version() | current_date |
+-----------+--------------+
| 8.0.25    | 2021-06-01   |
+-----------+--------------+
1 row in set (0.00 sec)

mysql>
```

这是一个简单的查询，说明了有关 MySQL 查询的几点信息：

（1）查询通常由以分号结尾的 SQL 语句组成（有一些例外可以省略分号，QUIT 语句就是其中之一）。

（2）当我们输入查询语句后，MySQL 将它发送到服务器执行并显示结果，然后显示出下一个"mysql>"提示符，表明它已准备好接收另一个查询。

（3）MySQL 以表格形式（行和列）显示查询的结果。第一行是列的标签（或称为列名），其他行是查询的结果。通常列标签是从数据库表中提取的列名称。如果要检索表达式的值而不是表列的值，那么 MySQL 会使用表达式本身来标记该列。

（4）MySQL 显示返回了多少行数据以及执行查询用了多长时间，由此我们大致可以了解服务器的性能。不过这些值是不精确的，因为它们代表的是时钟时间，并且受到服务器负载和网络延迟等因素的影响。

例 2.5　使用不同字母大小写进行相同内容的查询。

```
mysql> SELECT VERSION(), CURRENT_DATE;
mysql> select version(), current_date;
mysql> SeLeCt vErSiOn(), current_DATE;
```

查询结果如图 2-2 所示，在查询时，输入的 SQL 语句中的关键字可以不区分字母大小写，最后输出的

图 2-2　SQL 关键字不区分字母大小写示例

答案是一致的,也不会抛出任何异常。

例 2.6 把 MySQL 用作计算器。

```
mysql> select sin(pi()/5), (9+1)*5;
+--------------------+---------+
| sin(pi()/5)        | (9+1)*5 |
+--------------------+---------+
| 0.5877852522924731 |      50 |
+--------------------+---------+
1 row in set (0.00 sec)
```

执行结果如图 2-3 所示,MySQL 给出了计算结果。

例 2.7 用分号分隔多条 SQL 语句。

```
mysql> select version(); select now();
+-----------+
| version() |
+-----------+
| 8.0.25    |
+-----------+
1 row in set (0.00 sec)

+---------------------+
| now()               |
+---------------------+
| 2021-06-01 06:49:51 |
+---------------------+
1 row in set (0.00 sec)
```

执行结果如图 2-4 所示,可以在一行中输入多条 SQL 语句(或命令),每条 SQL 语句之间只需用分号分隔即可。

图 2-3　MySQL 作为计算器　　　　　　图 2-4　一行执行多条 SQL 语句

例 2.8 MySQL 可以写成多行的形式,最后带上终止分号即可。

```
mysql> select
    -> user()
    -> ,
    -> current_date;
```

```
+-----------------+--------------+
| user()          | current_date |
+-----------------+--------------+
| root@localhost  | 2021-06-01   |
+-----------------+--------------+
1 row in set (0.00 sec)
```

上述 SQL 语句的执行结果如图 2-5 所示。如果一条 SQL 查询语句冗长，在一行内写不下，那么可以写成多行的形式，最后带上终止分号即可，因为 MySQL 不是在每一行的末尾去找代表 SQL 语句结束的分号，而是在多个连续行去找分号来确定语句的结束位置。（MySQL 接收自由格式的输入，在看到分号之后才会执行当前的 SQL 查询语句），当我们不输入分号时，之前输入的命令是不会被执行的。

例 2.9　取消查询。

```
mysql> select
    -> users
    -> \c
mysql>
```

执行结果如图 2-6 所示，当输入"\c"之后，当前输入的命令就被取消执行了。

图 2-5　多行查询　　　　　图 2-6　取消执行当前输入的 SQL 命令

需要注意的是，在输入取消命令"\c"之后，MySQL 回到"mysql>"提示符状态，表示已准备好接收新的查询。表 2-11 总结了 MySQL 显示不同提示符时所处的状态。

表 2-11　MySQL 所处状态的含义

提　　示	意义和说明
mysql>	准备就绪，等待接收新的查询
->	等待多行查询的下一行
'>	等待下一行，等待以单引号（'）开头的字符串完成
">	等待下一行，等待以双引号（"）开头的字符串完成
`>	等待下一行，等待以反引号（`）开头的标识符完成
/*>	等待下一行，等待以"/*"开头的注释完成

（1）当以单行输入查询语句时，如果忘记以分号结尾，那么 MySQL 会一直等待我们输

入“;”：

```
mysql> select user()
    ->
```

如果发生这种情况（我们认为已经输入完成一条查询语句，而唯一的响应是"->"提示符），那么很可能就是 MySQL 在等待分号。输入分号之后，MySQL 才会开始执行输入的 SQL 查询语句：

```
mysql> select user()
    -> ;
+----------------+
| user()         |
+----------------+
| root@localhost |
+----------------+
```

（2）"'>"和"">"提示符表示当前正处于字符串的收集过程中（即表示 MySQL 正在等待用户输入表示字符串终止的配对符号）。在 MySQL 中，我们可以编写由一对"'"或一对"""作为起止的字符串，而 MySQL 允许输入跨多行的字符串。当看到"'>"或"">"提示符时，表示输入了包含以"'"或"""开头的字符串，但尚未输入终止字符串的配对引号。例如：

```
mysql> select * from mytable where name = 'clay and age <18;
    '>
```

如果输入此 select 语句，然后按 Enter 键并等待结果，则没有任何反应。此时我们需要注意"'>"提示符提供的线索。它其实是告诉我们：MySQL 希望看到字符串的其余部分（语句中 clay 开始的字符串缺少了第二个单引号）。此时我们可以选择输入"\c"取消本次查询语句的执行：

```
mysql> select * from mytable where name = clay and age < 18;
    '> '\c
mysql>
```

提示符又变回"mysql>"，表明 MySQL 再次就绪，可以接收新的查询。

2.3 创建数据库和表

在学习如何创建数据库和表之前，可以使用如下步骤来查询当前服务器中存在的数据库：

步骤 01 使用 show 语句查询当前服务器上存在哪些数据库：

```
mysql> show databases;
+--------------------+
| database           |
+--------------------+
| information_schema |
| mysql              |
| performance_schema |
```

```
| sys                |
+--------------------+
4 rows in set (0.00 sec)
```

从以上信息可以了解到，当我们执行 show databases 语句后，服务器将查询当前存在的所有数据库。

mysql 数据库存储了用户访问权限等信息。test 数据库是系统提供的默认测试库，可以删除。上述语句显示的数据库列表可能和读者计算机上显示的数据库列表有所不同，因为不会显示当前用户没有访问权限的数据库。

步骤02 如果 mysql 数据库存在，那么我们可以进入 mysql 数据库查询当前数据库中的表。比如，使用如下语句可以切换到 mysql 数据库：

```
mysql> use mysql
Database changed
```

使用如下语句可以查询当前数据库中的所有表：

```
mysql> show tables;
+---------------------------------------------+
| Tables_in_mysql                             |
+---------------------------------------------+
| columns_priv                                |
| component                                   |
| db                                          |
```

执行结果如图 2-7 所示，首先切换到 mysql 数据库，然后查询出当前 mysql 数据库中的所有表。

图 2-7　查询 mysql 数据库中的所有表

2.3.1　创建数据库

如果要创建一个新的数据库，那么可以执行如下语句：

```
create database DEMO;
```

执行结果如图 2-8 所示，创建数据库成功。

在 UNIX 操作系统中，数据库的名称是区分字母大小写的（这与 SQL 关键字不区分字母大小写不同），因此在进入数据库时，必须使用数据库的名称 DEMO 来指向正确的数据库，而不能使用 Demo、demo。对于表名也是如此（注意，在 Windows 中此限制不适用）。如图 2-9 所示，切换数据库时，由于未正确使用大小写而导致切换数据库操作失败。

如图 2-10 所示，唯有数据库名称的大小写正确才能正常完成数据库的切换。

图 2-8　创建数据库

图 2-9　数据库切换操作失败　　　　图 2-10　数据库切换操作成功

如果在创建数据库时出现诸如 ERROR 1044 (42000): Access denied for user clay@'localhost' 的错误，则说明用户没有创建数据库所需的权限。

2.3.2　创建表

如图 2-11 所示，进入 DMEO 数据库中，此时该数据库中并没有任何表。

如果要在 DEMO 数据库中创建表，那么在创建之前需要了解如下信息：

图 2-11　切换数据库并查询所有表

- 表名：代表数据表的名称，如果我们想要存储用户信息，则可以取名为 userinfo。
- 表字段名：表示表中有哪些属性，比如 name、age、address 等字段。

以下为创建 MySQL 数据表的通用语法：

```
create table table_name (column_name column_type);
```

例 2.10　在 DEMO 数据库中创建数据表 userinfo，SQL 语句如下：

```
create table if not exists `userinfo`(
  `id` int unsigned auto_increment,
  `name` varchar(100) not null,
  `age` int not null,
  `date` date,
  primary key ( `id` ))engine=innodb default charset=utf8;
```

上述语句说明如下：

- 如果不希望字段的值为 null，那么可以将字段的属性设置为 not null，如果在操作数据库时将 null 输入到该字段，则会报错。
- auto_increment：把列定义为自增的属性，一般用于主键，数值会自动加 1。
- primary key：用于把列定义为主键，可以使用多列来定义主键，列之间以逗号分隔。
- engine：设置存储引擎。
- charset：设置字符集的编码。
- varchar(100) 与 int：代表此字段使用的数据类型，后续章节会专门讲解每一种数据类型。

创建结果如图 2-12 所示。

图 2-12　创建表

如果想查询此表的结构，可以执行如下语句：

describe userinfo;

执行结果如图 2-13 所示，可以清晰地看到此表的结构及其说明。

图 2-13　查询表的结构及其说明

此后若再次查询当前数据库中的表，则会显示已创建表的基本信息，如图 2-14 所示。

图 2-14　查询数据库中表的基本信息

2.4　本章练习

1. 查看当前系统中所有的数据库。
2. 创建名为 userinfo 的数据库。
3. 根据表 2-5、表 2-6、表 2-8、表 2-9、表 2-10 的信息，使用 SQL 语句在数据库中创建表。

第 3 章

MySQL 数据类型

本章内容：

- 数字数据类型
- 日期和时间数据类型
- 字符串数据类型
- JSON 数据类型

本章将讲解 MySQL 提供的多种数据类型（包括 MySQL 新提供的数据类型）以及数据库对每种数据类型出现异常时的处理。

3.1 数字数据类型

在 MySQL 中，数字数据类型又分为如下 4 个类别：

- 整数类型：integer、int、smallint、tinyint、mediumint、bigint。
- 定点类型：decimal、numeric。
- 浮点类型：float、double。
- 位值类型：bit。

1. 整数类型

MySQL 支持 SQL 标准整数类型 int 和 smallint，也支持标准扩展的整数类型 tinyint、

mediumint 和 bigint。表 3-1 显示了每种整数类型所需的存储空间和取值范围，我们可以根据实际情况选用合适的整数类型。

表 3-1　整数类型所需的存储空间和取值范围

类　　型	存储空间（字节）	有符号的最小值	无符号的最小值	有符号的最大值	无符号的最大值
tinyint	1	–128	0	127	255
smallint	2	–32768	0	32767	65535
mediumint	3	–8388608	0	8388607	16777215
int	4	–2147483648	0	2147483647	4294967295
bigint	8	-2^{63}	0	$2^{63}-1$	$2^{64}-1$

2. 定点类型

decimal 和 numeric 类型用于保存对精度要求很高的数据，比如货币数据。在 MySQL 中，numeric 被实现为 decimal，所以适用于 decimal 数据类型的数据同样适用于 numeric 数据类型的数据。

在 decimal 列声明中，可以指定精度和小数位数。

例 3.1　在 decimal 列声明中指定精度为 5，小数位数为 2，SQL 语句如下：

salary decimal(5,2)

精度表示为值存储的有效位数，小数位数表示可以存储在小数点后的位数。decimal(5,2) 可以用于存储具有 5 位数字和两位小数的任何数值，因此定义为此数据类型的 salary 列的取值范围为–999.99～999.99。如果 decimal 小数位数为 0，则值不包含小数点或小数部分，但定义为 decimal 数据类型的列的实际取值范围可能会受到所定义的精度或小数位数的限制。如果给定义为 decimal 数据类型的列分配的值的小数点后的位数超过所允许的位数，那么数值会被截断到允许的位数。表 3-2 是 decimal 数据类型的详细说明。

表 3-2　decimal 数据类型的详细说明

类　　型	存储空间（字节）	有符号的最小值	无符号的最小值	有符号的最大值	无符号的最大值
decimal	对于 decimal(M,D)，如果 M>D，则为 M+2，否则为 D+2	取决于 M 和 D 的值	取决于 M 和 D 的值	取决于 M 和 D 的值	取决于 M 和 D 的值

3. 浮点类型

float 和 double 数据类型表示的是浮点数据类型。在 MySQL 中，单精度浮点数使用 4 字节，双精度浮点数使用 8 字节。表 3-3 是 float 和 double 数据类型的详细说明。

表 3-3　float 和 double 数据类型的详细说明

类　　型	存储空间（字节）	范围（有符号）	范围（无符号）
float	4	(−3.402 823 466 E+38，−1.175 494 351 E−38)，0，(1.175 494 351 E−38，3.402 823 466 351 E+38)	0，(1.175 494 351 E−38，3.402 823 466 E+38)
double	8	(−1.797 693 134 862 315 7 E+308，−2.225 073 858 507 201 4 E−308)，0，(2.225 073 858 507 201 4 E−308，1.797 693 134 862 315 7 E+308)	(2.225 073 858 507 201 4 E−308，1.797 693 134 862 315 7 E+308)

需要注意的是，如果要使程序具有最大的可移植性，那么在定义浮点数据的类型时，应该定义为 float 或 double precision 数据类型，且不要指定精度或位数。

4. 位值类型

即 bit 类型，它的表示方式为 bit(m)，m 的取值范围为 1～64。bit 类型存储的是二进制字符串。

bit 类型的数据范围为 bit(1)～bit(64)，换算成十进制，其值的取值范围为 $0 \sim 2^{64}-1$；tinyint unsigned 类型的数值取值范围为 0～255（十进制数）。bit 类型占用的存储空间近似为(m+7)/8 字节，而 tinyint 类型的存储空间为 1 字节。当使用 bit(1)和 tinyint 时，它们占用的存储空间是一致的。

5. 超出范围和溢出处理

当 MySQL 在数值列中存储超出列数据类型允许范围的值时，其结果取决于当时启用的 SQL 模式：

- 如果启用了严格的 SQL 模式，那么 MySQL 会根据 SQL 标准拒绝超出范围的值并显示错误提示信息，表明插入操作失败。
- 如果没有启用限制模式，那么 MySQL 会将插入的值截断为列数据类型范围的边界值来存储。
- 当把超出定义范围的值分配给定义为整数的列时，MySQL 将把列数据类型范围的边界值存储到列中。
- 当为浮点或定点类型的列分配的值超出指定（或默认）精度和小数位数所隐含的范围时，MySQL 将把列数据类型范围的边界值存储到列中。

1）超出范围处理

例 3.2　假设表 t1 具有以下定义：

```
create table t1 (i1 tinyint, i2 tinyint unsigned);
```

（1）启用严格 SQL 模式后，会出现超出范围的错误。

使用如下命令启用数据库的严格 SQL 模式：

```
mysql> set sql_mode = 'traditional';
```

在启用数据库的严格 SQL 模式下，插入超过数据类型范围的数值会被拒绝插入并抛出异常：

```
mysql> insert into t1 (i1, i2) values(256, 256);
ERROR 1264 (22003): Out of range value for column 'i1' at row 1
mysql> select * FROM t1;
Empty set (0.00 sec)
```

（2）如果未启用严格 SQL 模式，则会发出警告，但是不会报错。

使用如下语句设置未启用严格 SQL 模式：

```
mysql> set sql_mode = '';
```

如下语句是在未启用严格 SQL 模式的情况下插入超过数据类型范围的数值：

```
mysql> insert into t1 (i1, i2) values(256, 256);
```

如下语句是在查询上面的插入语句时出现的警告信息：

```
mysql>show warnings;
+---------+------+---------------------------------------------+
| Level   | Code | Message                                     |
+---------+------+---------------------------------------------+
| Warning | 1264 | Out of range value for column 'i1' at row 1 |
| Warning | 1264 | Out of range value for column 'i2' at row 1 |
+---------+------+---------------------------------------------+
```

如下语句用来检验当插入的数值超出数据类型的范围时，数值是否会插入数据表中：

```
mysql> select * from t1;
+------+------+
| i1   | i2   |
+------+------+
| 127  | 255  |
+------+------+
```

从结果可知，如果不启用严格 SQL 模式，在数值超过数据类型范围的情况下依然会把数据添加到数据表中，只不过发出了警告，同时实际插入的数值是数据类型范围的边界值，在本例中有边界，即为最大值。

2）溢出处理

数值表达式计算期间的溢出会导致错误。

例 3.3 有符号 bigint 值的最大值是 9223372036854775807，因此以下语句会产生错误：

```
mysql> select 9223372036854775807 + 1;
ERROR 1690 (22003): bigint value is out of range in '(9223372036854775807 + 1)'
```

在这种情况下，要想操作成功，可以将结果值转换为无符号数。

```
mysql> select cast(9223372036854775807 as unsigned) + 1;
```

```
+----------------------------------------+
| cast(9223372036854775807 as unsigned)+1 |
+----------------------------------------+
|              9223372036854775808        |
+----------------------------------------+
```

发生溢出的原因是结果值超过了数据类型范围的上限，所以处理上面这种错误的另一种方法就是修改数值的数据类型，例如把数值类型修改为浮点类型，因为浮点类型的取值范围大，对于此例不会出现异常情况：

```
mysql> select 9223372036854775807.0 + 1;
+--------------------------+
| 9223372036854775807.0 + 1 |
+--------------------------+
|     9223372036854775808.0 |
+--------------------------+
```

在默认情况下，unsigned 整数值之间的减法（其中一个类型为 unsigned）会产生一个无符号结果值。如果结果值是负数，则会导致错误：

```
mysql> set sql_mode = '';
Query OK, 0 rows affected (0.00 sec)
mysql> select cast(0 as unsigned) - 1;
ERROR 1690 (22003): bigint unsigned value is out of range in '(cast(0 as unsigned) - 1)'
```

如果 no_unsigned_subtraction 启用了 SQL 模式，则不会出现执行语句的异常：

```
mysql> set sql_mode = 'no_unsigned_subtraction';
mysql> select cast(0 as unsigned) - 1;
+-----------------------+
| cast(0 as unsigned) - 1 |
+-----------------------+
|                    -1 |
+-----------------------+
```

3.2 日期和时间数据类型

表示时间值的日期和时间类型有这样几种：datetime、date、timestamp、time 和 year。每种时间类型都有一个有效值范围和一个"零"值，当指定不符合规则的日期或时间数据时，MySQL 将使用"零"值来替换。

使用日期和时间类型时，需要注意以下事项：

- MySQL 以标准格式输出给定的日期或时间类型的值。如果使用日期或时间类型的格式不恰当，可能会出现不可预测的结果。
- MySQL 尝试以多种格式解释用户输入的值，但是日期部分必须按年-月-日的顺序给出（例如'21-06-01'），不能使用月-日-年或日-月-年的顺序（例如'01-06-21'、'06-01-21'）。

如果要将其他顺序的字符串转换为年-月-日的顺序，可以使用 str_to_date()函数进行转换。
- 对于包含2位数年份值的不明确日期，MySQL会使用以下规则进行解释：
  - ➢ 70—99范围内的年份值转换为1970—1999。
  - ➢ 00—69范围内的年份值转换为2000—2069。
- 默认情况下，当MySQL遇到无效的日期或时间类型的值时，会将该值转换为该类型的"零"值。如果时间类型的值超出范围，那么此值会被剪裁到时间范围的相应端点。
- MySQL允许将"零"值'0000-00-00'存储为"虚拟日期"。在某些情况下，这比使用null值更方便，并且使用更少的数据和索引空间。要禁止存储'0000-00-00'为"虚拟日期"，可以启用 no_zero_date 模式（set session sql_mode='strict_trans_tables,no_zero_date';）。

表3-4是所有日期和时间类型格式的详细说明。

表3-4 所有日期和时间类型格式的详细说明

类型	存储（字节）	范围	格式	用途
date	3	1000-01-01 到 9999-12-31	YYYY-MM-DD	日期值
time	3	'-838:59:59'到'838:59:59'	HH:MM:SS	时间值
year	1	1901 到 2155	YYYY	年份值
datetime	8	1000-01-01 00:00:00 到 9999-12-31 23:59:59	YYYY-MM-DD HH:MM:SS	日期和时间值
timestamp	4	1970-01-01 00:00:00 到 2038-1-19 11:14:07	YYYY-MM-DD HH:MM:SS	日期和时间值

3.3 字符串数据类型

在MySQL中，字符串数据类型有这样几种：char、varchar、text、binary、varbinary、blob、enum和set。对于数据类型定义为char、varchar和text的列，MySQL以字符为单位定义长度规范。对于数据类型定义为binary、varbinary和blob的列，MySQL以字节为单位定义长度规范。

当列定义为char、varchar、enum和set的数据类型时，同时还可以指定列的字符集，尤其在存储中文时，建议指定字符集格式为utf8，以防止出现乱码问题。

例3.4 创建表指定字段类型时设置字符集格式。创建一个名为c1的列，该列的字符集为utf8，以及一个名为c2的列，该列的字符集为latin1，SQL语句如下：

```
create table mytable
(
    c1 varchar(255) character set utf8,
```

```
    c2 text character set latin1 collate latin1_general_cs
);
```

1. char 和 varchar 类型

在 MySQL 中，char 和 varchar 类型很相似，但是它们被存储和检索的方式有所不同，而且在最大长度和是否保留尾随空格方面也不相同。

char 和 varchar 类型的声明需要规定存储的最大字符数。例如，char(50)最多可容纳 50 个字符。char 类型的列规定的长度可以是 0～255 的任何值。char 类型的列存储值时，会用空格向右填充到指定的长度。当设置启用 SQL 模式（即执行 "set sql_mode= 'pad_char_to_full_length';"）时，char 类型列被检索到的值，其尾随的空格不会被删除。

varchar 类型列中的值是可变长度的字符串。长度可以指定为 0～65535 的值。varchar 的有效最大长度受最大行大小和使用的字符集约束。与 char 值相比，varchar 值存储 1 字节或 2 字节的长度字段。如果内容字段不超过 255 字节，则长度字段使用 1 字节；如果内容长度超过 255 字节，则长度字段使用 2 字节。

如果未启用严格 SQL 模式并且为 char 或 varchar 类型的列分配的值超过该列的最大长度，则该值将被截断并生成警告信息。对于非空格字符的截断，可以使用严格的 SQL 模式提示错误（而不是警告）并禁止插入值。对于 varchar 类型的列，无论使用哪种 SQL 模式，超出列规定长度的尾随空格在插入之前都会被截断并生成警告信息。对于 char 列，无论使用哪种 SQL 模式，都会从插入的值中截断多余的尾随空格。

表 3-5 展示了 char 和 varchar 类型的差别（假设该列使用单字节字符集）。

表 3-5 char 和 varchar 类型的差别

内容	char(4)	占用字节	varchar(4)	占用字节
''	' '	4 字节	''	1 字节
'ab'	'ab '	4 字节	'ab'	3 字节
'abcd'	'abcd'	4 字节	'abcd'	5 字节
'abcdefgh'	'abcd'	4 字节	'abcd'	5 字节

如果给定的值存储在 char(4)和 varchar(4)列中，则从列中检索到的值并不总是相同的，因为 char 类型列检索时会从列中删除尾随空格。

例 3.5 说明 char 和 varchar 类型的差异的示例 SQL 语句及其执行结果如下：

```
mysql> create table tb(t varchar(4), b char(4));
query ok, 0 rows affected (0.01 sec)
mysql> insert into tb values ('ab ', 'ab ');
query ok, 1 row affected (0.00 sec)

mysql> select concat('[', t, ']'), concat('[', b, ']') from tb;
+---------------------+---------------------+
| concat('[', v, ']') | concat('[', c, ']') |
+---------------------+---------------------+
```

```
| [ab ]               | [ab]               |
+---------------------+--------------------+
1 row in set (0.03 sec)
```

2. binary 和 varbinary 类型

在MySQL中，binary和varbinary类型很相似。与char、varchar类型不同的是，它们存储的是字符串而不是字节串，而且binary和varbinary类型是以字节为单位来测量长度的，而不是以字符为单位。

当 binary 类型值被存储时，用右填充模式填充值到指定的长度，填充值是 0x00（0 字节）。也就是说当插入数据时，使用 0x00 向右填充，并且没有为检索删除尾随字节。比如，binary(3) 列插入'a'时变成'a \0\0'，插入'a\0'时变成'a\0\0'，两个插入的值在检索时保持不变。

例 3.6　binary 类型的值被存储时，填充值"0x00"影响列值存储的示例 SQL 语句及其执行结果如下：

```
mysql> create table t (c binary(3));
query ok, 0 rows affected (0.01 sec)

mysql> insert into t set c = 'a';
query ok, 1 row affected (0.01 sec)

mysql> select hex(c), c = 'a', c = 'a\0\0' from t;
+--------+---------+--------------+
| hex(c) | c = 'a' | c = 'a\0\0' |
+--------+---------+--------------+
| 610000 |       0 |            1 |
+--------+---------+--------------+
1 row in set (0.03 sec)
```

如上示例说明新增的值'a'最终被填充成'a\0\0'存储。

和 binary 类型不同的是，varbinary 类型列中的值是可变长度的字节串。在插入值时不会使用 0x00 填充，查询的时候也不会丢弃任何字节。在对列值进行比较的时候，所有的字节都是有效的，并且 0x00<space（space 对应的是 0x20）。

3. blob 和 text 类型

blob 类型的值是一个二进制的大对象，可以容纳可变数量的数据。tinyblob、blob、mediumblob 和 longblob 类型的区别仅在于它们可以存储的值的最大长度不相同。

blob 类型的值被视为二进制字符串（字节字符串），具有二进制字符集和排序规则，基于列值中字节的数值对值进行排序和比较。text 类型的值被视为非二进制字符串（字符串），具有除二进制之外的字符集，并且根据字符集的排序规则对值进行排序和比较。表3-6 是这几种类型数据的存储范围和使用用途说明。

表 3-6 字符串数据的存储范围和使用用途说明

类　　型	存　储　范　围	使　用　用　途
text	0～65535 字节	长文本数据
mediumblob	0～16777215 字节	二进制形式的中等长度文本数据
mediumtext	0～16777215 字节	中等长度文本数据
longblob	0～4294967295 字节	二进制形式的极大文本数据
longtext	0～4294967295 字节	极大文本数据

各个类型可存储的最大长度根据实际情况进行选择。

4. enum 类型

enum 类型（即枚举类型）的列值表示一个字符串对象，其值选自在表创建时列规范的枚举值。enum 类型具有以下优点：

- 在列具有有限的数据集合的情况下压缩数据空间。输入的字符串会自动编码为数字。
- 可读的查询和输出。在查询时，实际存储的数字被转换为相应字符串。

1）创建和使用 enum 列

枚举值必须是带引号的字符串。

例 3.7　创建和使用 enum 列的示例 SQL 语句及其执行结果如下：

```
mysql> create table mytable (
    ->     name varchar(40),
    ->     size enum('x-small', 'small', 'medium')
    -> );
query ok, 0 rows affected (0.02 sec)

mysql> insert into mytable (name, size) values ('','x-small'), ('','small'), ('',1);
query ok, 3 rows affected (0.01 sec)
records: 3  duplicates: 0  warnings: 0

mysql> select name, size from mytable;
+------+---------+
| name | size    |
+------+---------+
|      | x-small |
|      | small   |
|      | x-small |
+------+---------+
3 rows in set (0.00 sec)

mysql> update mytable set size = 'x-small' where size = 2;
query ok, 1 row affected (0.01 sec)

mysql> select name, size from mytable;
+------+---------+
| name | size    |
+------+---------+
```

```
|        |   x-small |
|        |   x-small |
|        |   x-small |
+--------+-----------+
3 rows in set (0.00 sec)
```

示例说明：在给字段使用 enum 类型的时候，MySQL 服务会自动给枚举的字符串添加索引编号，因此在操作过程中可以直接操作枚举值，也可以操作枚举索引编号。

使用枚举的时候，需要注意下面两个问题：

- 容易混淆。enum 类型的字段在底层其实是用整型来存储的，比如用 enum 类型来存储'3'、'2'、'1'这样的数字字符串时，其对应的索引编号是 1、2、3，不熟悉的开发人员很容易混淆。
- 字段值修改频繁的字段不适合使用 enum 类型。比如例 3.7 中的 size 字段，要增加一个尺寸，就要修改字段，给字段的 enum 类型增加一个尺寸枚举值。如果每增加一个枚举就要修改字段，这样维护成本较高，所以遇到这种情况不建议使用 enum 类型。

2）枚举文字的索引编号

每个枚举值都有一个索引编号，列规范中列出的元素都分配有索引编号，索引编号从 1 开始。空字符串的索引编号为 0，因此可以使用以下 select 语句查找 enum 类型列是否分配了无效值的行：

```
mysql> select * from tb_name where enum_col=0;
```

表 3-7 是枚举值和索引编号的规则说明。

表 3-7 枚举值和索引编号的规则说明

枚举字段内容	索 引 编 号
null	null
''	0
'男'	1
'女'	2
'未知'	3

使用 enum 类型列需要注意的是，最多只能有 65535 个不同的元素，即最多只能有 65535 个不同的枚举值。

如果要检索枚举的索引编号，可以通过如下语句去查询：

```
mysql>  select enum_col+0 from tbl_name;
```

5. set 类型

set 类型（集合类型）的列值表示可以有零个或多个字符串对象。一个 set 类型的列最多可以有 64 个不同的成员值，并且每个值必须从创建表时指定的值列表中选择。set 类型由多个逗号分隔的列值集合组成，set 类型成员值本身不应包含逗号。

例 3.8 指定 set('one', 'two') not null 的列可以具有以下任何值：

''
'one'
'two'
'one,two'

如果要将数字存储到 set 列中，则对于指定的列 set('a','b','c','d')，其成员应具有如表 3-8 所示的十进制值和二进制值。

表 3-8　set 类型字段值的十进制值和二进制值

set 类型字段值	十 进 制 值	二 进 制 值
'a'	1	0001
'b'	2	0010
'c'	4	0100
'd'	8	1000

无论插入值时元素的顺序是怎样的，查询展示时都是根据创建表指定的顺序列出的，而且无论插入值的次数是多少，该值中的每个元素都只会出现一次。

例 3.9 查询包含多个 set 元素的值，示例 SQL 语句如下：

```
# 一列被指定为set('a','b','c','d'):
mysql> create table myset (col set('a', 'b', 'c', 'd'));

# 如果插入值'a,d', 'd,a', 'a,d,d', 'a,d,a', 和'd,a,d'
mysql> insert into myset (col) values
    -> ('a,d'), ('d,a'), ('a,d,a'), ('a,d,d'), ('d,a,d');
Query OK, 5 rows affected (0.01 sec)
Records: 5  Duplicates: 0  Warnings: 0
```

最终查询的结果如下：

```
mysql> select col from myset;
+------+
| col  |
+------+
| a,d  |
| a,d  |
| a,d  |
| a,d  |
| a,d  |
+------+
5 rows in set (0.02 sec)
```

需要注意的是，当插入的值不为系统所支持时，该值将被忽略并发出警告。

例 3.10 往 set 列中插入的值不为系统所支持时，SQL 语句及其执行结果如下：

```
mysql> insert into myset (col) values ('a,d,d,s');
Query OK, 1 row affected, 1 warning (0.03 sec)
mysql> show warnings;
```

```
+---------+------+--------------------------------------------+
| Level   | Code | Message                                    |
+---------+------+--------------------------------------------+
| Warning | 1265 | Data truncated for column 'col' at row 1   |
+---------+------+--------------------------------------------+
1 row in set (0.04 sec)

mysql> select col FROM myset;
+------+
| col  |
+------+
| a,d  |
| a,d  |
| a,d  |
| a,d  |
| a,d  |
| a,d  |
+------+
6 rows in set (0.01 sec)
```

如上示例插入的值不被系统所支持时，系统发出了警告，只将支持的数据插入表中。

3.4 JSON 数据类型

MySQL 支持 JSON 数据类型，JSON 数据类型具有如下优势：

- 存储在 JSON 类型列中的 JSON 文档会被自动验证，无效的文档会产生错误。
- 存储在 JSON 类型列中的 JSON 文档会被转换为允许快速读取文档元素的内部格式。
- 在 MySQL 8 中，优化器可以执行 JSON 类型列的局部就地更新，而不用删除旧文档后再将整个新文档写入该列。

在 MySQL 中，JSON 类型列的值被写为字符串。如果字符串不符合 JSON 数据格式，则会产生错误。

例 3.11 插入无效的 JSON 值，示例 SQL 语句及其执行结果如下：

```
mysql> create table t1 (jdoc json);
Query OK, 0 rows affected (0.20 sec)

mysql> insert into t1 values('{"key1": "value1", "key2": "value2"}');
Query OK, 1 row affected (0.01 sec)

mysql> insert into t1 values('[1, 2,');
ERROR 3140 (22032) at line 2: Invalid JSON text:
"Invalid value." at position 6 in value (or column) '[1, 2,'.
```

从示例中可以看出，第 3 条语句插入的值明显不是 JSON 数据格式，所以抛出了异常。

1. JSON 操作函数

1）json_type()函数

用于返回对应的数据类型。

例 3.12 json_type()函数的使用，示例 SQL 语句及其执行结果如下：

```
mysql> select json_type('["a", "b","c" ]');
+-----------------------------+
| json_type('["a", "b", "c"]') |
+-----------------------------+
| ARRAY                       |
+-----------------------------+

mysql> select json_type('"a"');
+------------------+
| json_type('"a"') |
+------------------+
| STRinG           |
+------------------+

mysql> select json_type('a');
ERROR 3146 (22032): Invalid data type for JSON data in argument 1
to function json_type; a JSON string or JSON type is required.
```

由此可见，当值是 JSON 格式时，json_type 函数可以查询出该值的数据类型，当值不是 JSON 格式时，则会抛出异常。

2）json_array()函数（参数可以为空）

用于返回参数值的 JSON 数组。

例 3.13 json_array()函数的使用，示例 SQL 语句及其执行结果如下：

```
mysql> select json_array('a',false,1);
+-------------------------+
| json_array('a',false,1) |
+-------------------------+
| ["a", false, 1]         |
+-------------------------+
1 row in set (0.00 sec)

mysql> select json_array();
+--------------+
| json_array() |
+--------------+
| []           |
+--------------+
1 row in set (0.00 sec)
```

这两条查询分别用于将一个字符串"a"转换为 JSON 数组和获取空的 JSON 数组。

3）json_object 函数（参数可以为空）

用于返回键-值对（Key-Value Pair）的 JSON 对象。

例 3.14 json_object 函数的使用，示例 SQL 语句及其执行结果如下：

```
mysql> select json_object('key1','a','key2','b');
+------------------------------------+
| json_object('key1','a','key2','b') |
+------------------------------------+
| {"key1": "a", "key2": "b"}         |
+------------------------------------+
1 row in set (0.01 sec)
```

将字符串转换为 JSON 对象，其中包含两个键-值对，即"key1": "a"，"key2": "b"。

4）json_merge_preserve()函数

用于获取两个或多个 JSON 文档并返回组合结果。

例 3.15 json_merge_preserve()函数的使用，示例 SQL 语句及其执行结果如下：

```
mysql> select json_merge_preserve('["a", 123]', '{"key": "value"}');
+-------------------------------------------------------+
| json_merge_preserve('["a", 123]', '{"key": "value"}') |
+-------------------------------------------------------+
| ["a", 123, {"key": "value"}]                          |
+-------------------------------------------------------+
1 row in set (0.01 sec)
```

将两个 JSON 数组合并成一个 JSON 对象。

5）json_merge_patch()函数

用于返回多个 JSON 数据合并之后的对象。

例 3.16 json_merge_patch()函数的使用，示例 SQL 语句及其执行结果如下：

```
mysql> select json_merge_patch('{"a": 3, "b": 2}', '{"c": 3, "a": 4}', '{"c": 5, "d": 3}');
+-----------------------------------------------------------------------------+
| json_merge_patch('{"a": 3, "b": 2}', '{"c": 3, "a": 4}', '{"c": 5, "d": 3}')|
+-----------------------------------------------------------------------------+
| {"a": 4, "b": 2, "c": 5, "d": 3}                                            |
+-----------------------------------------------------------------------------+
1 row in set (0.00 sec)
```

将 3 个 JSON 对象合并为 1 个。

6）json_extract()函数

用于提取 JSON 对象值。

例 3.17 json_extract()函数的使用，示例 SQL 语句及其执行结果如下：

```
mysql> select json_extract('{"id": 18, "name": "clay"}', '$.name');
+------------------------------------------------------+
| json_extract('{"id": 18, "name": "clay"}', '$.name') |
+------------------------------------------------------+
| "clay"                                               |
+------------------------------------------------------+
```

```
1 row in set (0.00 sec)

mysql> select json_extract('{"id": 18, "name": "clay"}', '$.*');
+--------------------------------------------------+
| json_extract('{"id": 18, "name": "clay"}', '$.*') |
+--------------------------------------------------+
| [29, "Taylor"]                                   |
+--------------------------------------------------+
1 row in set (0.00 sec)
```

第一条查询用于提取 JSON 对象中"name"键对应的值。

第二条查询用于提取 JSON 对象中所有的键-值对。

7）json_extract()函数

用于提取 JSON 数组值。

例 3.18　json_extract()函数的使用，示例 SQL 语句及其执行结果如下：

```
// 提取单个元素
mysql> select json_extract('["a", "b", "c"]', '$[1]');
+----------------------------------------+
| json_extract('["a", "b", "c"]', '$[1]') |
+----------------------------------------+
| "b"                                    |
+----------------------------------------+
1 row in set (0.00 sec)

// 范围提取
mysql> select json_extract('["a", "b", "c"]', '$[1 to 2]');
+---------------------------------------------+
| json_extract('["a", "b", "c"]', '$[1 to 2]') |
+---------------------------------------------+
| ["b", "c"]                                  |
+---------------------------------------------+
1 row in set (0.00 sec)
```

8）json_replace()函数

用于替换值。

例 3.19　json_replace()函数的使用，示例 SQL 语句及其执行结果如下：

```
mysql> select json_replace('{"id": 18, "name": "clay"}', '$.name', 'clay2');
+--------------------------------------------------------------+
| json_replace('{"id": 18, "name": "clay"}', '$.name', 'clay2') |
+--------------------------------------------------------------+
| {"id": 18, "name": "clay2"}                                  |
+--------------------------------------------------------------+
1 row in set (0.00 sec)
```

将 JSON 字符串中名为"name"的属性值替换为"clay2"。

9）json_set()函数

用于设置值（替换旧值，并插入不存在的新值）。

例 3.20 json_set()函数的使用，示例 SQL 语句及其执行结果如下：

```
#设置的值为不存在的新值,则插入
mysql> select json_set('{"id": 18, "name": "clay"}', '$.name2', 'clay2');
+------------------------------------------------------------+
| json_set('{"id": 18, "name": "clay"}', '$.name2', 'clay2') |
+------------------------------------------------------------+
| {"id": 18, "name2'": "clay2", "name": "clay"}              |
+------------------------------------------------------------+
1 row in set (0.00 sec)
```

10）json_insert()函数

用于插入值（插入新值，但不替换已经存在的旧值）。

例 3.21 json_insert()函数的使用，示例 SQL 语句及其执行结果如下：

```
// 把元素值插入指定的下标位置，如果之前该下标位置不存在元素值，则插入成功
mysql> select json_insert('[1, 2, 3]', '$[4]', 4);
+-------------------------------------+
| json_insert('[1, 2, 3]', '$[4]', 4) |
+-------------------------------------+
| [1, 2, 3, 4]                        |
+-------------------------------------+
1 row in set (0.00 sec)

// 把元素值插入指定的下标位置，如果该下标位置处已经存在元素值，则插入失败
mysql> select json_insert('[1, 2, 3]', '$[1]', 4);
+-------------------------------------+
| json_insert('[1, 2, 3]', '$[1]', 4) |
+-------------------------------------+
| [1, 2, 3]                           |
+-------------------------------------+
1 row in set (0.00 sec)
```

11）json_remove()函数

删除 JSON 数据，删除指定值后的 JSON 文档。

例 3.22 json_remove()函数的使用，示例 SQL 语句及其执行结果如下：

```
// 删除指定下标位置处的数据
 mysql> select json_remove('[1, 2, 3]', '$[1]');
+----------------------------------+
| json_remove('[1, 2, 3]', '$[1]') |
+----------------------------------+
| [1, 3]                           |
+----------------------------------+
1 row in set (0.00 sec)

// 根据属性名称删除数据
mysql> select json_remove('{"id": 18, "name": "clay"}', '$.name');
+-----------------------------------------------------+
| json_remove('{"id": 18, "name": "clay"}', '$.name') |
+-----------------------------------------------------+
| {"id": 18}                                          |
+-----------------------------------------------------+
```

```
1 row in set (0.00 sec)
```

2. JSON 值和非 JSON 值转换

在 MySQL 中操作 JSON 类型时，除了前面给出的各种函数之外，在实际开发中经常还会出现 JSON 类型和其他数据类型转换的情况，表 3-9 是 JSON 类型和其他数据类型转换时遵循的规则说明。

表3-9　JSON类型和其他数据类型转换遵循的规则说明

数 据 类 型	CAST（其他类型转换为 JSON）	CAST（JSON 转换为其他类型）
JSON	没有变化	没有变化
utf8 字符类型（utf8mb4、utf8、ascii）	该字符串被解析为 JSON 值	JSON 值被序列化为 utf8mb4 字符串
其他字符类型	其他编码的字符被隐式转换为 utf8 编码并按照 utf8 编码进行处理	JSON 值被序列化为 utf8mb4 字符串，然后转换为其他字符编码
null	结果 null 转换为 JSON 类型的值	不适用

3.5　本章练习

1. 创建一张用户信息表，要求性别字段的数据类型为 enum 类型。
2. 说出 float 和 double 数据类型的区别。

第 2 部分　MySQL 基本操作

第 4 章　MySQL 基础操作
第 5 章　MySQL 连接查询和数据复制
第 6 章　MySQL 基础函数

第 4 章

MySQL 基础操作

本章内容：

- MySQL 数据库操作
- MySQL 数据表操作
- MySQL 数据查询
- 在 MySQL 中插入数据
- MySQL 中的 where、update、delete 语句
- MySQL 中的常见函数、分组等操作

本章将讲解 MySQL 创建数据库、创建数据表、数据查询、常用函数和分组以及一些常用的操作语句，并且提供大量实例供读者参考和学习。

4.1 数据库操作

数据库常见的操作有：创建数据库、删除数据库和修改数据库。

1. 创建数据库

创建数据库的语法如下：

```
create database [if not exists] db_name
```

其中，db_name 代表的是需要创建的数据库的名称；[if not exists]是可选的，表示如果此名称的数据库不存在，就创建一个此名称的数据库，防止因创建现存同名的数据库而报错。

例 4.1 两次创建名为 mydb 的数据库，示例 SQL 语句及其执行结果如下：

```
mysql> create database if not exists mydb;
Query OK, 1 row affected, 1 warning (0.00 sec)

mysql> create database  mydb;
ERROR 1007 (HY000): Can't create database 'mydb'; database exists
mysql> show databases;
+--------------------+
| Database           |
+--------------------+
| information_schema |
| mydb               |
| mysql              |
| performance_schema |
| sys                |
+--------------------+
5 rows in set (0.02 sec)
```

如上所示，在第一次成功创建 mydb 数据库之后，第二次创建同名的数据库时，由于没有加 if not exists 条件而报错，说明此名称的数据库已经存在。

2. 删除数据库

删除数据库的语法如下：

```
drop database [if exists] db_name;
```

其中，db_name 代表的是需要删除的数据库的名称；[if exists]是可选的，表示如果此名称的数据库存在，就删除此名称的数据库，防止因删除不存在的数据库而报错。

例 4.2 两次删除名为 mydb 的数据库，示例 SQL 语句及其执行结果如下：

```
mysql> create database  mydb;
Query OK, 1 row affected (0.00 sec)

mysql> drop database mydb;
Query OK, 0 rows affected (0.00 sec)

mysql> drop database mydb;
ERROR 1008 (HY000): Can't drop database 'mydb'; database doesn't exist
mysql> drop database if exists  mydb;
Query OK, 0 rows affected, 1 warning (0.00 sec)

mysql> show databases;
+--------------------+
| Database           |
+--------------------+
| information_schema |
| mysql              |
| performance_schema |
| sys                |
+--------------------+
4 rows in set (0.00 sec)
```

如上所示，在第一次成功删除 mydb 数据库后，第二次删除 mydb 数据库时，由于没有加 if exists 条件而报错，说明已不存在 mydb 数据库。

3．修改数据库

修改数据库的语法如下：

```
alter database db_name
[ default character set charset_name]
[ default collate collation_name]
```

其中 db_name 代表的是需要修改的数据库的名称；charset_name（字符集名）表示数据库默认使用的字符集编码；collation_name（排序规则）表示数据库默认使用的排序规则。这条语句就是用于修改 db_name 数据库的默认字符集和默认排序规则。

例 4.3 修改数据库的默认字符集编码，示例 SQL 语句及其执行结果如下：

```
mysql> create database mydb;
Query OK, 1 row affected (0.01 sec)

mysql> alter database mydb character set=utf8;
Query OK, 1 row affected, 1 warning (0.00 sec)
```

在创建数据库时，如果想修改数据库的默认字符集（即字符集编码），则可以通过上面的命令进行修改。

4.2　数据表操作

数据表常见的操作有创建数据表、修改数据表和删除数据表。

4.2.1　创建数据表

创建数据表的语法如下：

```
create table [if not exists] tbl_name (
    字段 1 数据类型 [字段属性|约束][索引][注释],
    ...
    字段 n 数据类型 [字段属性|约束][索引][注释]
)[表类型][表字符集][注释];
```

其中，tbl_name 表示数据表的名称；[if not exists]是可选的，表示当不存在此名称的数据表时才去创建，防止因创建已经存在的数据表而报错。

例 4.4 创建数据表，示例 SQL 语句及其执行结果如下：

```
mysql> create database if not exists mydb;
Query OK, 1 row affected, 1 warning (0.00 sec)
```

```
mysql> use mydb
Database changed
mysql>
mysql> create table if not exists `use_tbl`(
    ->     `use_id` int unsigned auto_increment,
    ->     `use_title` varchar(100) not null,
    ->     `use_author` varchar(40) not null,
    ->     `submission_date` date,
    ->     primary key ( `use_id` )
    -> )engine=innodb default charset=utf8;
Query OK, 0 rows affected, 2 warnings (0.00 sec)
```

其中，use_id 字段使用了 auto_increment（自动增加），表示此字段的值默认从 1 开始，每次自动加 1。下面列举创建表时的其他组成部分。

1. 字段指定为主键

主键（Primary Key）是一行数据的唯一标识，而且此键的列必须定义为 not null。如果它们没有显式地声明为 not null，则 MySQL 会隐式地声明为 not null。一张数据表只能有一个主键。

例 4.5 创建一张 student 表，其中 studentno 字段为此数据表的主键，示例 SQL 语句如下：

```
create table student (
    `studentno` int(4)  primary key,
    ...);
```

2. 给字段和数据表添加注释

例 4.6 创建一张包含 id 字段的数据表 test，给表和字段分别添加注释，示例 SQL 语句如下：

```
create table test (
    `id` int(11) unsigned comment 'this is num'
)comment='test table ';
```

3. 为数据表设置字符集编码

语法如下：

```
create table [if not exists] 表名(
    # 省略代码
) charset = 字符集名;
```

4. 指定存储引擎

语法如下：

```
create table 表名(
    # 省略代码
)engine=存储引擎;
```

为了更好地理解上述语法，下面用一个实例来说明。

例 4.7 根据表 4-1 的说明，创建一个数据表结构。

表 4-1 创建数据表结构

字段名称	字段说明	数据类型	长度	属性
studentno	学号	int	4	非空\主键
loginpwd	密码	varchar	20	非空
studentname	姓名	varchar	50	非空
sex	性别	char	2	非空，默认为"男"
gradeid	年级编号	int	4	无符号
phone	电话	varchar	50	
address	地址	varchar	255	默认值为"地址不详"
borndate	出生日期	datetime		
email	邮件账户	varchar	50	
identitycard	身份证号码	varchar	18	唯一

示例 SQL 语句如下：

```
create table `student`(
    `studentno` int(4) not null comment '学号' primary key,
    `loginpwd` varchar(20) not null comment '密码',
    `studentname` varchar(50) not null comment '学生姓名',
    `sex` char(2) default '男' not null comment '性别',
    `gradeid` int(4) unsigned comment '年级编号',
    `phone` varchar(50) comment '联系电话',
    `address` varchar(255) default '地址不详'comment '地址', `borndate` datetime comment '出生时间',
    `email` varchar(50) comment'邮件账户',
    `identitycard` varchar(18) unique key comment '身份证号码'
) comment='学生表';
```

4.2.2 数据表的修改与删除

除了创建数据表之外，还可以修改和删除数据表。

1. 修改数据表

下面是修改数据表的一些常见语法。

（1）修改表名的语法如下：

```
alter table 旧表名 rename [to] 新表名;
```

（2）添加字段的语法如下：

```
alter table 表名 add 字段名 数据类型 [属性];
```

（3）修改字段的语法如下：

```
alter table 表名 change 原字段名 新字段名 数据类型 [属性];
```

(4) 删除字段的语法如下：

```
alter table 表名 drop 字段名;
```

例 4.8 结合修改数据表的常见语法进行测试，示例 SQL 语句如下：

```
drop table if exists `demo01`;
# 创建表
create table `demo01`(
   `id` int(10) not null auto_increment primary key,
   `name` varchar(8) not null
);
# 修改表名
alter table `demo01` rename `demo02`;
# 添加字段
alter table ` demo02 ` add `password` varchar(32) not null;
# 修改字段
alter table ` demo02 ` change `name` `username`char(10) not null;
# 删除字段
alter table ` demo02 ` drop `password`;
```

2．删除数据表

删除 MySQL 数据表的通用语法如下：

```
drop table [if exists] table_name ;
```

其中，table_name 代表的是需要删除的数据表的名称；[if exists]是可选的，表示如果此名称的数据表存在，就删除此名称的数据表，防止因删除不存在的数据表而报错。

提　示

在 MySQL 中删除表时要小心，因为执行删除命令后所有数据都会消失。

4.2.3　约束

约束（Constraint）是 MySQL 提供的自动保持数据库完整性的一种方法，定义了可输入表或表的单个列中的数据的限制条件。在 MySQL 中有 5 种约束，具体如下：

（1）主键约束（Primary Key Constraint）：要求主键的列数据唯一，并且不允许为空。

（2）唯一约束（Unique Constraint）：要求该列的数据唯一，允许为空，但只能出现一个空值。

（3）检查约束（Check Constraint）：某列取值的范围限制、格式限制等。

（4）默认约束（Default Constraint）：某列的默认值。

（5）外键约束（Foreign Key Constraint）：用于在两个数据表之间建立关系，需要指定引用主表的哪一列。

可以在创建表时规定约束（通过 create table 语句），或者在表创建之后规定约束（通过 alter table 语句）。

创建约束的语法如下：

```
alter table 表名  add constraint 约束名
```

创建约束时需要注意约束名的取名规则，推荐采用下面的命名方式：

- 主键约束：如 pk_stuno。
- 唯一约束：如 uq_stuid。
- 默认约束：如 df_stuaddress。
- 检查约束：如 ck_stuborndate。
- 外键约束：如 fk_stuno。

下面我们通过一个实例来加深对创建约束的理解。

例 4.9 在表 student 中添加下面的约束：

（1）主键约束：学号。

（2）唯一约束：身份证号码。

（3）默认约束：address 列的默认值是"地址不详"。

（4）检查约束：出生日期的值为 1980 年 1 月 1 日后。

（5）外键约束：设置 gradeid 列为外键，建立 grade 表与 student 表的引用关系。

创建约束的示例 SQL 语句及其执行结果如下：

```
# --主键约束
alter table student
add constraint pk_stuno primary key (studentno);
# --唯一约束（身份证号码唯一）
alter table student add
constraint uq_stuid unique (identitycard);
# --默认约束（地址不详）
alter table studentadd
constraint df_stuaddress default ('地址不详') for address;
# --检查约束（出生日期是自 1980 年 1 月 1 日以后）
alter table student
add constraint ck_stuborndate check(borndate>='1980-1-1');
# --添加外键约束，需要先在 grade 表中给 gradeid 创建主键，然后在 student 表中给 gradeid 创建外键
alter table grade
add constraint pk_gradeid primary key(gradeid);
alter table student
add constraint fk_grade
foreign key(gradeid) references grade(gradeid);
```

需要特别强调的是，创建数据表与创建数据库等语句都属于数据定义语言（Data Definition Language，DDL），MySQL 8 支持原子操作数据定义语句。原子 DDL 语句将与操作 DDL 关

联的数据字典更新、存储引擎操作和二进制日志写入合并为一个原子操作,即使服务器在操作期间停止,该操作也会被提交(更改将保留到数据字典中)或被回滚。在早期的 MySQL 版本中,元数据存储在元数据文件、非事务表和特定于存储引擎的字典中,是需要中间提交的。而 MySQL 8 的数据字典提供的集中式事务元数据存储解决了这些问题,让 DDL 语句操作重构为原子操作(之前的版本不属于原子操作)。

4.3 数据查询

MySQL 数据库使用 select 语句来查询数据,其语法如下:

```
select column_name,column_name
from table_name[where clause][limit n][ offset m]
```

在查询之前需要掌握下面几个知识点:

(1) select 语句中可以使用一个或者多个数据表,表之间使用逗号(,)分隔,并使用 where 语句来设定查询条件。

(2) select 命令可以读取一条或者多条记录。

(3) 可以使用星号(*)来代替其他字段,select 语句会返回表的所有字段数据。

(4) 可以使用 where 语句来包含任何条件。

(5) 可以使用 limit 属性来设置返回的记录数。

(6) 可以通过 offset 指定 select 语句开始查询的数据偏移量。默认情况下数据偏移量为 0。

例 4.10 使用 select 语句进行查询,示例 SQL 语句及其执行结果如下:

```
# 查询 userinfo 表中所有的数据
mysql> select * from userinfo;
+----+----------+------+
| id | name     | age  |
+----+----------+------+
|  2 | zhangsan |   27 |
|  3 | lisi2    |   27 |
+----+----------+------+
2 rows in set (0.00 sec)
 # 查询 userinfo 表中年龄大于 18 并且姓名是 zhangsan 的数据
mysql> select * from userinfo where age='18' and name="zhangsan";
Empty set (0.00 sec)

 # 在 userinfo 表中从 1 开始,连续查询一条数据
mysql> select name,age  from userinfo limit 1,1;
+-------+------+
| name  | age  |
+-------+------+
| lisi2 |   27 |
+-------+------+
1 row in set (0.00 sec)
```

4.4 数据插入

使用 insert into 语句在 MySQL 数据表中插入数据，其语法如下：

```
insert into table_name ( field1, field2,...fieldn )
                values
                    ( value1, value2,...valuen );
```

需要注意的是，如果数据是字符串型，则必须使用单引号或者双引号，如"value"，而且插入的数据值和字段类型必须对应，并且数据表中的列数必须与要插入的列数相匹配。

例 4.11　向数据表 userinfo 中插入两条数据，示例 SQL 语句及其执行结果如下：

```
mysql> use mydb;
Database changed
mysql> create table userinfo(
    ->     id Int(10) not null auto_increment primary key,
    -> name varchar(20),
    -> age int(2)
    -> );
Query OK, 0 rows affected, 2 warnings (0.02 sec)
# 插入数据
mysql> insert into userinfo
    ->        (name,age)
    -> values
    ->        ('zhangsan',26)
    -> ;
Query OK, 1 row affected (0.00 sec)
# 插入数据
mysql> insert into userinfo
    ->        (name,age)
    -> values
    ->        ('lisi',26);
Query OK, 1 row affected (0.00 sec)
# 查询所有数据，验证数据是否全部插入
mysql> select * from userinfo;
+----+----------+------+
| id | name     | age  |
+----+----------+------+
|  1 | zhangsan |   26 |
|  2 | lisi     |   26 |
+----+----------+------+
2 rows in set (0.00 sec)
```

在这个示例中，我们并没有提供 id 的数据，因为该字段在创建表时已经设置为 auto_increment（自动增加），所以该字段会自动递增，而不需要我们去设置数据值。

如果需要在一行中插入多个数据，可以使用如下语法：

```
insert into table_name ( field1, field2,...fieldn )
                values
                    ( value1, value2,...valuen ),
                    (value1, value2,...valuen);
```

4.5　where（筛选）

通常情况下，在 MySQL 数据表中可以使用 SQL select 语句来读取数据。如果需要有条件地从表中选取数据，那么可以将 where 子句添加到 select 语句中，其语法如下：

```
select field1, field2,...fieldn from table_name1, table_name2...
[where condition1 [and [or]] condition2...
```

在使用此语法时，需要了解如下几个知识点：

（1）查询语句中可以使用一张或者多张数据表，表之间使用逗号（,）分隔，并使用 where 语句来设置查询条件。

（2）可以在 where 子句中指定任何条件。

（3）可以使用 and 或者 or 指定一个或多个条件。

（4）where 子句也可以运用于 SQL 的 delete 或者 update 命令中。

（5）where 子句类似于程序语言中的 if 条件，根据 MySQL 表中的字段值来读取指定的数据。

另外，where 子句可以使用如表 4-2 所示的运算符。

表 4-2　where 子句可以使用的运算符

运算符	说明
=	等号，检测两个值是否相等，如果相等则返回 true
<>、!=	不等于，检测两个值是否相等，如果不相等则返回 true
>	大于号，检测左边的值是否大于右边的值，如果左边的值大于右边的值则返回 true
<	小于号，检测左边的值是否小于右边的值，如果左边的值小于右边的值则返回 true
>=	大于等于号，检测左边的值是否大于或等于右边的值，如果左边的值大于或等于右边的值则返回 true
<=	小于等于号，检测左边的值是否小于或等于右边的值，如果左边的值小于或等于右边的值则返回 true

例 4.12　使用 where 子句在 MySQL 表中查询指定数据，示例 SQL 语句及其执行结果如下：

```
# 查询表中所有数据
mysql> select * from userinfo;
+----+----------+------+
| id | name     | age  |
+----+----------+------+
|  1 | zhangsan |   26 |
|  2 | lisi     |   26 |
+----+----------+------+
2 rows in set (0.00 sec)
```

```
# 查询表中的数据，必须满足 1=1（相当于恒成立）
mysql> select * from userinfo where 1=1;
+----+----------+------+
| id | name     | age  |
+----+----------+------+
|  1 | zhangsan |   26 |
|  2 | lisi     |   26 |
+----+----------+------+
2 rows in set (0.00 sec)

# 查询表中 age 字段值大于 18 的数据
mysql> select * from userinfo where age>18;
+----+----------+------+
| id | name     | age  |
+----+----------+------+
|  1 | zhangsan |   26 |
|  2 | lisi     |   26 |
+----+----------+------+
2 rows in set (0.00 sec)
# 查询表中 name 字段等于 'lisi' 的数据
mysql> select * from userinfo where name='lisi';
+----+------+------+
| id | name | age  |
+----+------+------+
|  2 | lisi |   26 |
+----+------+------+
1 row in set (0.00 sec)

# 查询表中 name 字段不等于 'lisi' 的数据
mysql> select * from userinfo where name!='lisi';
+----+----------+------+
| id | name     | age  |
+----+----------+------+
|  1 | zhangsan |   26 |
+----+----------+------+
1 row in set (0.01 sec)

# 查询表中 age 字段小于或等于 18 的数据
mysql> select * from userinfo where age<=18;
Empty set (0.00 sec)
```

如果想在 MySQL 数据表中读取指定的数据，那么 where 子句是非常有用的。如果给定的条件在表中没有任何匹配的记录，那么查询不会返回任何数据。

4.6　and 和 or（与和或）

and 和 or 运算符基于一个以上的条件对记录进行筛选。and 为"与"运算符，只有满足所有条件的数据才会被筛选出来。or 为"或"运算符，满足其中一个条件的数据就会被筛选出来。

例 4.13 使用 and 和 or 运算符查询数据，示例 SQL 语句及其执行结果如下：

```
mysql> select * from userinfo;
+------+-------+------+
| id   | name  | age  |
+------+-------+------+
|    1 | lisi  |   18 |
|    2 | lisi2 |   19 |
|    3 | lisi3 |   19 |
+------+-------+------+
3 rows in set (0.00 sec)

# 查询满足 id 大于 2 并且年龄大于或等于 19 的数据
mysql> select * from userinfo where id>2 and age>=19;
+------+-------+------+
| id   | name  | age  |
+------+-------+------+
|    3 | lisi3 |   19 |
+------+-------+------+
1 row in set (0.00 sec)

# 查询满足 id 大于 2 或者年龄大于或等于 19 的数据
mysql> select * from userinfo where id>2 or age>=19;
+------+-------+------+
| id   | name  | age  |
+------+-------+------+
|    2 | lisi2 |   19 |
|    3 | lisi3 |   19 |
+------+-------+------+
2 rows in set (0.00 sec)
```

4.7　between（范围查找）

between 运算符用于筛选出介于两个值之间（数据范围内）的值。

例 4.14 使用 between 运算符查询数据，示例 SQL 语句及其执行结果如下：

```
# 查询年龄在 18 和 19 岁之间的数据，包括 18 和 19
mysql> select * from userinfo where age between 18 and 19;
+------+-------+------+
| id   | name  | age  |
+------+-------+------+
|    1 | lisi  |   18 |
|    2 | lisi2 |   19 |
|    3 | lisi3 |   19 |
+------+-------+------+
3 rows in set (0.00 sec)
# 如下示例等同于上述使用 between 的语句，查询年龄在 18 和 19 岁之间的数据，包括 18 和 19
mysql> select * from userinfo where age>=18 and age<= 19;
+------+-------+------+
| id   | name  | age  |
+------+-------+------+
```

```
|   1 | lisi  |  18 |
|   2 | lisi2 |  19 |
|   3 | lisi3 |  19 |
+------+-------+------+
3 rows in set (0.00 sec)
```

4.8 distinct（去重关键字）

在 select 语句中，可以使用 distinct 关键字指示 MySQL 消除重复的记录值。

例 4.15 使用与不使用 distinct 显示查询结果，示例 SQL 语句及其执行结果如下：

```
# 去重显示 userinfo 表中的 age 列数据
mysql> select distinct age from userinfo ;
+------+
| age  |
+------+
|  18  |
|  19  |
+------+
2 rows in set (0.00 sec)
# 不去重显示 userinfo 表中的 age 列数据
mysql> select age from userinfo ;
+------+
| age  |
+------+
|  18  |
|  19  |
|  19  |
+------+
3 rows in set (0.00 sec)
```

4.9 update（修改或更新）

update 命令用于修改或更新 MySQL 数据表中的数据。使用 update 命令修改 MySQL 数据表中数据的通用语法如下：

```
update table_name set field1=new-value1, field2=new-value2[where clause]
```

在使用 update 命令时需要知道下面几个知识点：

（1）update 同时更新一个或多个字段。
（2）update 可以在 where 子句中指定任何条件。
（3）update 可以在单个数据表中同时更新数据。

例 4.16 使用 update 更新数据，示例 SQL 语句及其执行结果如下：

```
mysql> select * from userinfo where name='lisi';
```

```
+----+------+------+
| id | name | age  |
+----+------+------+
|  2 | lisi |   26 |
+----+------+------+
1 row in set (0.00 sec)

# 把 name 等于'lisi'的数据行中的 name 字段值修改为'lisi2'
mysql> update userinfo set name='lisi2' where name='lisi';
Query OK, 1 row affected (0.00 sec)
Rows matched: 1  Changed: 1  Warnings: 0

# 验证是否修改成功
mysql> select * from userinfo;
+----+----------+------+
| id | name     | age  |
+----+----------+------+
|  1 | zhangsan |   26 |
|  2 | lisi2    |   26 |
+----+----------+------+
2 rows in set (0.00 sec)

# 给所有用户的 age 值加 1
mysql> update userinfo set age=age+1;
Query OK, 2 rows affected (0.00 sec)
Rows matched: 2  Changed: 2  Warnings: 0

# 验证是否修改成功
mysql> select * from userinfo;
+----+----------+------+
| id | name     | age  |
+----+----------+------+
|  1 | zhangsan |   27 |
|  2 | lisi2    |   27 |
+----+----------+------+
2 rows in set (0.00 sec)
```

除此之外，如果 update 语句包含 order by（排序）子句，则各行会按照子句指定的顺序更新。

例 4.17　使用 order by 按指定的列排序，示例 SQL 语句及其执行结果如下：

```
mysql> select * from userinfo;
+----+----------+------+
| id | name     | age  |
+----+----------+------+
|  1 | zhangsan |   27 |
|  2 | lisi2    |   27 |
+----+----------+------+
2 rows in set (0.00 sec)

# 修改用户 id，在之前的基础上加 1
mysql> update userinfo set id=id+1;
ERROR 1062 (23000): Duplicate entry '2' for key 'userinfo.PRIMARY'
```

```
# 修改用户 id，在之前的基础上加 1，先排序，再依次修改
mysql> update userinfo set id=id+1 order by id desc;
Query OK, 2 rows affected (0.01 sec)
Rows matched: 2  Changed: 2  Warnings: 0
```

上面示例中数据表 userinfo 的主键是 id，由于第一个 update 更新时没有指定排序，因此在更新时出现错误（主键重复），重新指定排序之后可以正常进行修改。

4.10 delete（删除）

MySQL 数据库使用 delete 命令来删除 MySQL 数据表中的记录，通用语法如下：

```
delete from table_name [where clause]
```

在使用 delete 命令时需要知道下面几个知识点：

（1）如果没有指定 where 子句，则 MySQL 表中的所有记录都将被删除。
（2）可以在 where 子句中指定任何条件。
（3）可以在单张表中一次性删除所有数据。

例 4.18　使用 delete 命令删除数据，示例 SQL 语句及其执行结果如下：

```
mysql> select * from userinfo;
+----+----------+------+
| id | name     | age  |
+----+----------+------+
|  2 | zhangsan |   27 |
|  3 | lisi2    |   27 |
+----+----------+------+
2 rows in set (0.00 sec)

# 删除年龄等于 18 的数据
mysql> delete from userinfo where age=18;
Query OK, 0 rows affected (0.00 sec)

# 删除 name 等于 lisi2 的数据
mysql> delete from userinfo where name='lisi2';
Query OK, 1 row affected (0.00 sec)

# 验证是否删除成功
mysql> select * from userinfo ;
+----+----------+------+
| id | name     | age  |
+----+----------+------+
|  2 | zhangsan |   27 |
+----+----------+------+
1 row in set (0.01 sec)

# 删除当前表中的所有数据
mysql> delete from userinfo ;
```

```
Query OK, 1 row affected (0.00 sec)

mysql> select * from userinfo ;
Empty set (0.00 sec)
```

4.11　like（模糊匹配）

在 MySQL 中使用 select 命令来读取数据，同时可以在 select 语句中使用 where 子句来获取指定的记录。where 子句中可以使用等号（=）来设置筛选数据的条件，如"title='www.baidu.com'"。有时需要筛选出 title 字段中含有"com"字符的所有记录，这时就需要在 where 子句中使用 like 子句。

like 子句中使用百分号（%）字符来表示任意字符，如果没有使用百分号，那么 like 子句与等号（=）的效果是一样的。以下是 select 语句使用 like 子句从数据表中读取数据的通用语法：

```
select field1, field2,...fieldn
from table_name
where field1 [not] like condition1 [and [or]] filed2 = 'somevalue'
```

在使用 like 子句时需要知道下面几个知识点：

（1）可以在 where 子句中指定任何条件。

（2）可以在 where 子句中使用 like 子句。

（3）可以使用 like 子句代替等号（=）。

（4）like 通常与%一起使用，类似于一个元字符的搜索。

（5）可以使用 and 或者 or 运算符指定一个或多个条件。

（6）可以在 delete 或 update 命令中使用 where…like 子句来指定条件。

例 4.19～例 4.22 是进行 like 操作的示例。

例 4.19　在数据表 students 中查找所有以字母"H"开头的学生姓名，示例 SQL 语句及其执行结果如下：

```
mysql> select name from students
    -> where name like 'H%';
+--------+
| name   |
+--------+
| Hhoma  |
| Hom    |
+--------+
2 rows in set (0.12 sec)
```

可以看到，查询结果中只返回了以字母"H"开头的学生姓名。

例 4.20　在数据表 students 中查找所有不以字母 "F" 开头的学生姓名，示例 SQL 语句及其执行结果如下：

```
mysql> select name from tb_students_info
    -> where name not like 'F%';
+-------+
| name  |
+-------+
| Danys |
| Greey |
| Henry |
| Janen |
| Jimk  |
+-------+
5 rows in set (0.00 sec)
```

可以看到，查询结果中返回了不以字母 "F" 开头的学生姓名。

例 4.21　在数据表 students 中查找所有包含字母 "e" 的学生姓名，示例 SQL 语句及其执行结果如下：

```
mysql> select name from tb_students_info
    -> where name like '%e%';
+-------+
| name  |
+-------+
| Greey |
| Henry |
| Janen |
+-------+
2 rows in set (0.00 sec)
```

可以看到，查询结果中返回了所有包含字母 "e" 的学生姓名。

"_" 通配符只能代表单个字符，字符的长度不能为 0。例如，a_b 可以代表 acb、adb、aub 等字符串。

例 4.22　在数据表 students 中，查找所有以字母 "y" 结尾且 "y" 前面只有 4 个字母的学生姓名，示例 SQL 语句及其执行结果如下：

```
mysql> select name from students
    -> where name like '____y';
+-------+
| name  |
+-------+
| Henry |
+-------+
1 row in set (0.00 sec)
```

可以看到，查询结果中返回了以字母 "y" 结尾且 "y" 前面只有 4 个字母的学生姓名。

4.12　order by（排序）

使用 select 语句从 MySQL 数据表中读取数据后，如果需要对读取的数据进行排序，就可以使用 order by 子句来设置按照哪个字段、哪种方式进行排序，再返回搜索结果，通用的语法如下：

```
select field1, field2,...fieldn from table_name1, table_name2...
order by field1 [asc [desc][默认 asc]], [field2...] [asc [desc][默认 asc]]
```

在使用 order by 子句之前，需要了解下面的几个知识点：

（1）可以使用任何字段作为排序的条件，从而返回排序后的查询结果。

（2）可以设置多个字段来排序。

（3）可以使用 asc 或 desc 关键字来设置查询结果按升序或降序排列。默认情况下，按升序排列（asc）。

（4）可以添加 where…like 子句来设置条件。

例 4.23　通过 order by 对数据进行升序和降序，示例 SQL 语句及其执行结果如下：

```
mysql> use mydb;
Database changed
# 升序查询
mysql> select * from tbl order by submission_date asc;
+-------+-------------+-----------+-----------------+
| id    | title       | author    | submission_date |
+-------+-------------+-----------+-----------------+
| 3     | C#          | clay1     | 2015-05-01      |
| 4     | Java        | clay2     | 2016-03-06      |
| 1     | PHP         | clay3     | 2017-04-12      |
| 2     | MySQL       | clay4     | 2017-04-12      |
+-------+-------------+-----------+-----------------+
4 rows in set (0.01 sec)

# 降序查询
mysql> select * from tbl order by submission_date desc;
+----------+---------+---------+-----------------+
| id       | title   | author  | submission_date |
+----------+---------+---------+-----------------+----
| 1        | PHP     | clay3   | 2017-04-12      |
| 2        | MySQL   | clay4   | 2017-04-12      |
| 4        | Java    | clay2   | 2016-03-06      |
| 3        | C#      | clay1   | 2015-05-01      |
+----------+---------+---------+-----------------+
4 rows in set (0.01 sec)
# 带筛选条件的降序查询
mysql> select * from tbl where submission_date>='2016-03-06'
    order by submission_date desc;
+----------+---------+---------+-----------------+
| id       | title   | author  | submission_date |
+----------+---------+---------+-----------------+----
```

```
| 1        | PHP      | clay3    | 2017-04-12       |
| 2        | MySQL    | clay4    | 2017-04-12       |
| 4        | Java     | clay2    | 2016-03-06       |
+----------+----------+----------+------------------+
4 rows in set (0.01 sec)
```

4.13　group by（分组）

group by 语句根据一个或多个列对结果集进行分组。在分组的列上可以使用 count、sum、avg 等函数来统计结果。分组的语法如下：

```
select column_name, function(column_name)
from table_name
where column_name operator value
group by column_name;
```

例 4.24　使用 group by 语句将数据表按名字进行分组，示例 SQL 语句及其执行结果如下：

```
# 准备数据
# 因为内容中有中文，所以需要把客户端操作格式设置为utf8
set names utf8;
set foreign_key_checks = 0;
drop table if exists `employee_tbl`;
create table `employee_tbl` (
  `id` int(11) not null,
  `name` char(10) not null default '',
  `date` datetime not null,
  `singin` tinyint(4) not null default '0' comment '登录次数',
  primary key (`id`)
) engine=innodb default charset=utf8;
# 插入测试数据
insert into `employee_tbl` values ('1', 'clay', '2016-04-22 15:25:33', '1'), ('2', 'alsi', '2016-04-20 15:25:47', '3'), ('3', 'tom', '2016-04-19 15:26:02', '2'), ('4', 'alsi', '2016-04-07 15:26:14', '4'), ('5', 'clay', '2016-04-11 15:26:40', '4'), ('6', 'clay', '2016-04-04 15:26:54', '2');
# 验证数据是否插入成功
mysql> select * from employee_tbl;
+----+--------+---------------------+--------+
| id | name   | date                | singin |
+----+--------+---------------------+--------+
| 1  | clay   | 2016-04-22 15:25:33 |   1    |
| 2  | alsi   | 2016-04-20 15:25:47 |   3    |
| 3  | tom    | 2016-04-19 15:26:02 |   2    |
| 4  | alsi   | 2016-04-07 15:26:14 |   4    |
| 5  | clay   | 2016-04-11 15:26:40 |   4    |
| 6  | clay   | 2016-04-04 15:26:54 |   2    |
+----+--------+---------------------+--------+
6 rows in set (0.00 sec)
# 使用 group by 语句将数据表按名字进行分组，并统计每个人有多少条记录
mysql> select name, count(*) from  employee_tbl group by name;
+--------+----------+
| name   | count(*) |
+--------+----------+
```

```
| tom    |         1 |
| clay   |         3 |
| alsi   |         2 |
+--------+-----------+
3 rows in set (0.01 sec)
```

4.14　with rollup（分组统计再统计）

如果想在分组统计数据的基础上再进行相同的统计（比如 sum、avg、count 等），那么需要使用 with rollup 来实现，其语法如下：

```
select column_name, function(column_name)
from table_name
where column_name operator value
group by column_name with rollup;
```

例 4.25　将例 4.24 中的数据表按名字进行分组，再统计每个人登录的次数，示例 SQL 语句及其执行结果如下：

```
mysql> select name, sum(singin) as singin_count  from  employee_tbl group by name with rollup;
+--------+--------------+
| name   | singin_count |
+--------+--------------+
| tom    |            2 |
| clay   |            7 |
| alsi   |            7 |
| null   |           16 |
+--------+--------------+
4 rows in set (0.00 sec)
```

其中记录 null 表示所有人的登录次数。使用 coalesce 设置一个可以取代 null 的名称，示例 SQL 语句及其运行结果如下：

```
mysql> select coalesce(name, '总数'), sum(singin) as singin_count from  employee_tbl group by name with rollup;
+--------------------------+--------------+
| coalesce(name, '总数')   | singin_count |
+--------------------------+--------------+
| tom                      |            2 |
| clay                     |            7 |
| alsi                     |            7 |
| 总数                     |           16 |
+--------------------------+--------------+
4 rows in set (0.01 sec)
```

4.15　having（分组筛选）

由于 where 关键字无法与聚合函数一起使用，因此 MySQL 中增加了 having 子句，让我们可以筛选出分组后的各组数据。having 子句的语法如下：

```
select column_name, aggregate_function(column_name)
from table_name where column_name operator value
group by column_name
having aggregate_function(column_name) operator value;
```

例 4.26　根据 name 分组求出登录总次数，然后在计算结果的基础上统计出总次数大于 2 次的结果，示例 SQL 语句及其运行结果如下：

```
mysql> select name, sum(singin) as singin_count from  employee_tbl group by name having sum(singin)>2;
+------+--------------+
| name | singin_count |
+------+--------------+
| clay |            7 |
| alsi |            7 |
+------+--------------+
2 rows in set (0.00 sec)
```

4.16　union 和 union all（组合查询）

union 运算符将两个以上的 select 语句的结果连接组合到一个结果集合中，遇到多个 select 语句时会删除重复的数据。union 的语法如下：

```
select expression1, expression2, ... ,expression_n
from tables
[where conditions]
union [all | distinct]
select expression1, expression2, ... ,expression_n
from tables
[where conditions];
```

参数说明如下：

- expression1, expression2, …, expression_n：要检索的列。
- tables：要检索的数据表。
- where conditions：可选，检索条件。
- distinct：可选，删除结果集合中重复的数据。默认情况下，union 运算符已经删除了重复数据，所以 distinct 修饰符对结果没什么影响。
- all：可选，返回所有结果集，包含重复数据。

下面用示例来演示 union 和 union all 的使用。首先准备数据：

```
# 数据准备
mysql> drop table if exists userinfo;
Query OK, 0 rows affected (0.00 sec)

mysql> create table userinfo(id int,name varchar(20),age int);
Query OK, 0 rows affected (0.01 sec)
```

```
mysql> insert into userinfo(id,name,age) values(1,'lisi',18);
Query OK, 1 row affected (0.00 sec)

mysql> insert into userinfo(id,name,age) values(2,'lisi2',19);
Query OK, 1 row affected (0.01 sec)

mysql> insert into userinfo(id,name,age) values(3,'lisi3',19);
Query OK, 1 row affected (0.01 sec)

mysql> select * from userinfo;
+------+-------+------+
| id   | name  | age  |
+------+-------+------+
|  1   | lisi  |  18  |
|  2   | lisi2 |  19  |
|  3   | lisi3 |  19  |
+------+-------+------+
3 rows in set (0.00 sec)

mysql> drop table if exists student;
Query OK, 0 rows affected, 1 warning (0.00 sec)

mysql>create table student(id int,name varchar(20),age int,address varchar(50));
Query OK, 0 rows affected (0.01 sec)

mysql>insert into student(id,name,age,address ) values(1,'lisi',18,'shanghai');
Query OK, 1 row affected (0.00 sec)

mysql> insert into student(id,name,age,address ) values(2,'wangwu',19,'wuhan');
Query OK, 1 row affected (0.01 sec)

mysql>insert into student(id,name,age,address ) values(3,'zhangsan',20,'xian');
Query OK, 1 row affected (0.00 sec)

mysql> select * from student;
+------+----------+------+----------+
| id   | name     | age  | address  |
+------+----------+------+----------+
|  1   | lisi     |  18  | shanghai |
|  2   | wangwu   |  19  | wuhan    |
|  3   | zhangsan |  20  | xian     |
+------+----------+------+----------+
3 rows in set (0.00 sec)
```

例 4.27 使用 union 从数据表 userinfo 和 student 中选取所有不同的 age（只有不同的值），示例 SQL 语句及其执行结果如下：

```
 mysql> select age from student
    -> union
    -> select age from userinfo
    -> order by age ;
+------+
| age  |
+------+
```

```
|   18 |
|   19 |
|   20 |
+------+
3 rows in set (0.00 sec)
```

由此可知，使用 union 进行组合查询可以得到不重复的值。

例 4.28　使用 union all 从数据表 userinfo 和 student 中选取所有的 age（有重复的值），示例 SQL 语句及其执行结果如下：

```
mysql> select age from student
    ->     union all
    ->     select age from userinfo
    ->     order by age ;
+------+
| age  |
+------+
|   18 |
|   18 |
|   19 |
|   19 |
|   19 |
|   20 |
+------+
6 rows in set (0.00 sec)
```

由此可知，使用 union all 进行组合查询得到的值没有去重。

例 4.29　使用带有 where 的 union all 语句从数据表 userinfo 和 student 中选取所有大于或等于 18 的 age，示例 SQL 语句及其执行结果如下：

```
mysql> select age from student where address='shanghai'
    ->     union all
    ->     select age from userinfo where age>=18
    ->     order by age ;
+------+
| age  |
+------+
|   18 |
|   19 |
|   19 |
+------+
3 rows in set (0.00 sec)
```

4.17　本章练习

建立一个图书馆管理系统的数据库用来存放图书馆的相关信息，包括图书的基本信息、图书的借阅信息和读者的信息。根据下表信息使用 SQL 语句创建相应的表结构。

图书信息表 Book

列 名 称	数 据 类 型	说　明
BID	字符	图书编号，主键，该栏必填，必须以"ISBN"开头
BName	字符	图书书名，该栏必填
Author	字符	作者姓名
PubComp	字符	出版社
PubDate	日期	出版日期，必须小于当前日期
BCount	整数	现存数量，必须大于或等于 1
Price	货币	单价，必须大于 0

读者信息表 Reader

列 名 称	数 据 类 型	说　明
RID	字符	读者编号，主键，该栏必填
RName	字符	读者姓名，该栏必填
LendNum	整数	已借书数量，必须大于或等于 0
RAddress	字符	联系地址

图书借阅表 Borrow

列 名 称	数 据 类 型	说　明
RID	字符	读者编号，复合主键，读者信息表的外键，该栏必填
BID	字符	图书编号，复合主键，图书信息表的外键，该栏必填
LendDate	日期	借阅日期，复合主键，默认值为当前日期，该栏必填
WillDate	日期	应归还日期，必须大于或等于借阅日期，默认值为借阅日期+1 个月
ReturnDate	日期	实际归还日期，默认值为空

罚款记录表

列 名 称	数 据 类 型	说　明
RID	字符	读者编号，复合主键，读者信息表的外键，该栏必填
BID	字符	图书编号，复合主键，图书信息表的外键，该栏必填
PDate	日期	罚款日期，复合主键，默认值为当前日期，该栏必填
PType	整型	罚款类型，1-延期 2-损坏 3-丢失
Amount	货币	罚款金额，必须大于 0

1. 根据上述表格中的说明列，为每个表的相关列添加约束。
2. 在图书信息表中添加 BTotal 列，数据类型是 int，用于保存每种图书的馆藏总量。
3. 向各表中插入至少两条测试数据，并查询测试。

第 5 章

MySQL 连接查询和数据复制

本章内容：
- 内连接
- 左连接
- 右连接
- 复制表和临时表的使用

连接是关系数据库模型的主要特点。连接查询是关系数据库中主要的查询方式，包括内连接、外连接等。通过连接运算符可以实现多表查询。在关系数据库管理系统中，表创建时各数据之间的关系不必确定，常把一个实体的所有信息存放在一张表中。当查询数据时，通过连接操作查询出存放在多张表中的不同实体的信息。当两张或多张表中存在相同意义的字段时，便可以通过这些字段对不同的表进行连接查询。本章将介绍多表之间的内连接查询、外连接查询以及数据复制。

5.1 连接查询分类

前面章节讲述的都是从一张数据表中读取数据，但是在真正应用中经常需要从多张数据表中读取数据。接下来将向读者介绍如何使用 MySQL 的 join 运算符在两张或多张数据表中查询数据；可以在 select、update 和 delete 语句中使用 join 来联合多表查询。

join 按照功能大致分为如下 3 类：

（1）inner join（内连接）：获取两张表中字段匹配关系的记录。

（2）left join（左连接）：获取左表所有记录，即使右表没有对应匹配的记录。

（3）right join（右连接）：获取右表所有记录，即使左表中没有对应匹配的记录。

在理解这 3 种模式之前，先准备好下面两张表的数据。数据脚本如下：

```
drop table if exists a;
create table a (aid int, anum varchar(100));
insert into a(aid,anum) values(1,'a20210111');
insert into a(aid,anum) values(2,'a20210112');
insert into a(aid,anum) values(3,'a20210113');
insert into a(aid,anum) values(4,'a20210114');
insert into a(aid,anum) values(5,'a20210115');
drop table if exists b;
create table b (bid int, bnum varchar(100));
insert into b(bid,bnum) values(1,'a20220111');
insert into b(bid,bnum) values(2,'a20220112');
insert into b(bid,bnum) values(3,'a20220113');
insert into b(bid,bnum) values(4,'a20220114');
insert into b(bid,bnum) values(8,'a20220115');

mysql> select * from a;
+------+-----------+
| aid  | anum      |
+------+-----------+
|    1 | a20210111 |
|    2 | a20210112 |
|    3 | a20210113 |
|    4 | a20210114 |
|    5 | a20210115 |
+------+-----------+
5 rows in set (0.01 sec)

mysql> select * from b;
+------+-----------+
| bid  | bnum      |
+------+-----------+
|    1 | a20220111 |
|    2 | a20220112 |
|    3 | a20220113 |
|    4 | a20220114 |
|    8 | a20220115 |
+------+-----------+
5 rows in set (0.00 sec)
```

5.2　inner join（内连接）

inner join（内连接）用于获取两张表中满足条件的交集，如图 5-1 所示。

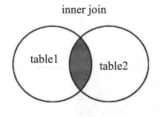

图 5-1　内连接

例 5.1　使用 MySQL 的 inner join（也可以省略 inner，只使用 join，效果一样）来连接 a

表 a 和 b 表，通过 aid 和 bid 的连接条件查询内连接的结果，示例 SQL 语句及其查询结果如下：

```
mysql> select * from a inner join b on a.aid=b.bid;
+------+-----------+------+-----------+
| aid  | anum      | bid  | bnum      |
+------+-----------+------+-----------+
|    1 | a20210111 |    1 | a20220111 |
|    2 | a20210112 |    2 | a20220112 |
|    3 | a20210113 |    3 | a20220113 |
|    4 | a20210114 |    4 | a20220114 |
+------+-----------+------+-----------+
4 rows in set (0.00 sec)
mysql> select * from a join b on a.aid=b.bid;
+------+-----------+------+-----------+
| aid  | anum      | bid  | bnum      |
+------+-----------+------+-----------+
|    1 | a20210111 |    1 | a20220111 |
|    2 | a20210112 |    2 | a20220112 |
|    3 | a20210113 |    3 | a20220113 |
|    4 | a20210114 |    4 | a20220114 |
+------+-----------+------+-----------+
4 rows in set (0.01 sec)

mysql> select * from a inner join b on a.aid=b.bid where aid>2;
+------+-----------+------+-----------+
| aid  | anum      | bid  | bnum      |
+------+-----------+------+-----------+
|    3 | a20210113 |    3 | a20220113 |
|    4 | a20210114 |    4 | a20220114 |
+------+-----------+------+-----------+
2 rows in set (0.00 sec)
```

第一条查询语句使用 inner join 将 a 表和 b 表连接起来，并且返回了两张表中 aid=bid 的所有记录；第二条查询语句与第一条查询语句等效，只不过使用了"join"而不是"inner join"；第三条查询语句进行了条件筛选，只返回 aid>2 的记录，即只返回 aid 为 3 和 4 的记录，并且这些记录还符合 inner join 条件（即 aid=bid）。

5.3　left join（左连接）

left join 会读取左边的数据表的全部数据，即使右边的数据表无对应数据，如图 5-2 所示。

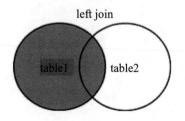

图 5-2　左连接

例 5.2 使用 MySQL 的 left join 来连接 a 表和 b 表，通过 aid 和 bid 的连接条件查询左连接的结果，示例 SQL 语句及其查询结果如下：

```
mysql> select * from a left join b on a.aid=b.bid;
+------+-----------+------+-----------+
| aid  | anum      | bid  | bnum      |
+------+-----------+------+-----------+
|    1 | a20210111 |    1 | a20220111 |
|    2 | a20210112 |    2 | a20220112 |
|    3 | a20210113 |    3 | a20220113 |
|    4 | a20210114 |    4 | a20220114 |
|    5 | a20210115 | null | null      |
+------+-----------+------+-----------+
5 rows in set (0.01 sec)

mysql> select * from a left join b on a.aid=b.bid where aid>2;
+------+-----------+------+-----------+
| aid  | anum      | bid  | bnum      |
+------+-----------+------+-----------+
|    3 | a20210113 |    3 | a20220113 |
|    4 | a20210114 |    4 | a20220114 |
|    5 | a20210115 | null | null      |
+------+-----------+------+-----------+
3 rows in set (0.00 sec)
```

第一条查询语句使用了左连接（left join）将 a 表和 b 表连接起来，以 a 表的 aid 列作为连接条件查看所有行，即使在连接条件中没有匹配的行也会显示在结果集中；第二条查询语句在第一条查询语句的基础上添加了 where 子句，过滤掉 aid 小于或等于 2 的行。因此，第二条查询语句只返回 aid 为 3、4 和 5 的行。

5.4 right join（右连接）

MySQL 的 right join 会读取右边的数据表的全部数据，即使左边的数据表无对应数据，如图 5-3 所示。

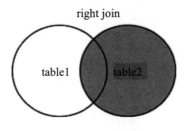

图 5-3 右连接

例 5.3 使用 MySQL 的 right join 来连接 a 表和 b 表，通过 aid 和 bid 的连接条件查询右连接的结果，示例 SQL 语句及其查询结果如下：

```
mysql> select * from a right join b on a.aid=b.bid;
```

```
+------+-----------+------+-----------+
| aid  | anum      | bid  | bnum      |
+------+-----------+------+-----------+
|    1 | a20210111 |    1 | a20220111 |
|    2 | a20210112 |    2 | a20220112 |
|    3 | a20210113 |    3 | a20220113 |
|    4 | a20210114 |    4 | a20220114 |
| null | null      |    8 | a20220115 |
+------+-----------+------+-----------+
5 rows in set (0.00 sec)

mysql> select * from a right join b on a.aid=b.bid where aid>2;
+------+-----------+------+-----------+
| aid  | anum      | bid  | bnum      |
+------+-----------+------+-----------+
|    3 | a20210113 |    3 | a20220113 |
|    4 | a20210114 |    4 | a20220114 |
+------+-----------+------+-----------+
2 rows in set (0.00 sec)
```

第一条查询语句使用了右连接（right join）将 a 表和 b 表连接起来，以 b 表的 bid 列作为连接条件查看所有行，即使在连接条件中没有匹配的行也会显示在结果集中，同时保留 b 表中没有匹配的行。因此，第一个查询结果中，最后一行显示了一个 null 值，因为 b 表中有一个 id 值为 8，但是在 a 表中没有对应的匹配行。

第二条查询语句在第一条查询语句的基础上添加了 where 子句，过滤掉 aid 小于或等于 2 的行，因此，第二条查询语句只返回 aid 为 3 和 4 的行。

5.5 in 和 not in（多关键字筛选）

in 运算符允许在 where 子句中规定多个值，而 not in 运算符则表示查询除了规定之外的所有结果集。

例 5.4　演示在 MySQL 中如何使用 in 和 not in 运算符，示例 SQL 语句及其执行结果如下：

```
mysql> select * from a ;
+------+-----------+
| aid  | anum      |
+------+-----------+
|    1 | a20210111 |
|    2 | a20210112 |
|    3 | a20210113 |
|    4 | a20210114 |
|    5 | a20210115 |
+------+-----------+
5 rows in set (0.00 sec)
# 查询 aid 包含 1、2、3 的结果集
mysql> select * from a where aid in(1,2,3);
+------+-----------+
| aid  | anum      |
+------+-----------+
```

```
|    1 | a20210111  |
|    2 | a20210112  |
|    3 | a20210113  |
+------+------------+
3 rows in set (0.00 sec)
# 查询 aid 不包含 1、2、3 的结果集
mysql> select * from a where aid not in(1,2,3);
+------+------------+
| aid  | anum       |
+------+------------+
|    4 | a20210114  |
|    5 | a20210115  |
+------+------------+
2 rows in set (0.00 sec)
# 查询 b 表中 bid 的所有结果
mysql> select bid from b;
+------+
| bid  |
+------+
|    1 |
|    2 |
|    3 |
|    4 |
|    8 |
+------+
5 rows in set (0.00 sec)
```

in 和 not in 中也可以是一个子查询。

例 5.5　in 和 not in 中是一个子查询，示例 SQL 语句及其查询结果如下：

```
mysql> select * from a where aid not in(select bid from b);
+------+------------+
| aid  | anum       |
+------+------------+
|    5 | a20210115  |
+------+------------+
1 row in set (0.00 sec)
```

这条 MySQL 查询语句用于从 a 表中选择 aid 不等于 b 表中 bid 的记录。具体来说，该查询语句使用 not in 关键字来筛选出 a 表中不满足条件的记录，其中 b 表中的 bid 字段被用作一个子查询。

5.6　exists 和 not exists（是否返回结果集）

exists 运算符用于判断查询子句是否有记录，如果有一条或多条记录，则返回 true，否则返回 false。而 not exists 与 exists 的结果相反。

例 5.6　演示在 MySQL 中如何使用 exists 和 not exists 运算符，示例 SQL 语句及其查询结果如下：

```
# 查询 a 表中的 aid 和 b 表中的 bid 匹配的结果集
mysql> select * from a where exists (select * from b where a.aid=b.bid);
+------+-----------+
| aid  | anum      |
+------+-----------+
|    1 | a20210111 |
|    2 | a20210112 |
|    3 | a20210113 |
|    4 | a20210114 |
+------+-----------+
4 rows in set (0.00 sec)
# 查询 a 表中的 aid 和 b 表中的 bid 不匹配的结果集
mysql> select * from a where not exists (select * from b where a.aid=b.bid);
+------+-----------+
| aid  | anum      |
+------+-----------+
|    5 | a20210115 |
+------+-----------+
1 row in set (0.00 sec)
```

5.7 复制表

复制表分为只复制表数据和完全复制表（包括表的结构、索引、默认值）。如果想要把一张表的数据复制到不同结构的表中，通常选择只复制表数据；如果想要备份表或者克隆相同的一张表，就选择完全复制表。只复制表数据需要提前创建一张表，而完全复制表不需要再创建表。

例 5.7　完全复制表，示例 SQL 语句及其执行结果如下：

```
# 不需要创建表，直接复制表结构和数据
mysql> select * from userinfo;
+------+-------+------+
| id   | name  | age  |
+------+-------+------+
|    1 | lisi  |   18 |
|    2 | lisi2 |   19 |
|    3 | lisi3 |   19 |
+------+-------+------+
3 rows in set (0.00 sec)

mysql> create table userinfo2 select id,name,age from userinfo;
Query OK, 3 rows affected (0.01 sec)
Records: 3  Duplicates: 0  Warnings: 0

mysql> select * from userinfo2;
+------+-------+------+
| id   | name  | age  |
+------+-------+------+
|    1 | lisi  |   18 |
|    2 | lisi2 |   19 |
|    3 | lisi3 |   19 |
+------+-------+------+
```

3 rows in set (0.00 sec)

例 5.8 创建表，然后只复制表数据，示例 SQL 语句及其执行结果如下：

```
# 查询要复制表的数据创建脚本
mysql> show create table userinfo \G;
*************************** 1. row ***************************
       Table: userinfo
Create Table: create table `userinfo` (
  `id` int default null,
  `name` varchar(20) default null,
  `age` int default null
) engine=InnoDB default charset=utf8mb3
1 row in set (0.00 sec)
# 根据复制表的结构创建表
 create table `userinfo3` (
    -> `id` int default null,
    -> `name` varchar(20) default null,
    -> `age` int default null
    -> ) engine=InnoDB default charset=utf8mb3
    -> ;
# 只复制表数据
mysql> insert into userinfo3 select * from userinfo;
Query OK, 3 rows affected (0.01 sec)
Records: 3  Duplicates: 0  Warnings: 0
mysql> select * from userinfo3;
+------+-------+------+
| id   | name  | age  |
+------+-------+------+
|    1 | lisi  |   18 |
|    2 | lisi2 |   19 |
|    3 | lisi3 |   19 |
+------+-------+------+
3 rows in set (0.00 sec)
```

5.8 临 时 表

MySQL 临时表用于保存一些临时数据。临时表只在当前连接可见，一旦关闭连接，MySQL 会自动删除临时表并释放所有空间。

例 5.9 演示临时表在 MySQL 中的状态，示例 SQL 语句及其执行结果如下：

```
# 创建临时表
mysql> create temporary table `tempuserinfo` (
    -> `id` int default null,
    -> `name` varchar(20) default null,
    -> `age` int default null
    -> ) engine=innodb default charset=utf8mb3;
Query OK, 0 rows affected, 1 warning (0.01 sec)
# 把 userinfo 数据复制到临时表 tempuserinfo
mysql> insert into tempuserinfo select * from userinfo;
Query OK, 3 rows affected (0.00 sec)
Records: 3  Duplicates: 0  Warnings: 0
# 查询临时表 tempuserinfo 中的数据
```

```
mysql> select * from tempuserinfo;
+------+-------+------+
| id   | name  | age  |
+------+-------+------+
|    1 | lisi  |  18  |
|    2 | lisi2 |  19  |
|    3 | lisi3 |  19  |
+------+-------+------+
3 rows in set (0.00 sec)
# 当查询所有表名时，临时表是查询不出来的
mysql> show tables;
+----------------+
| Tables_in_mydb |
+----------------+
| employee_tbl   |
| student        |
| test           |
| use_tbl        |
| userinfo       |
| userinfo2      |
| userinfo3      |
+----------------+
7 rows in set (0.00 sec)
# 退出当前登录的 MySQL
mysql> exit
Bye
# 重新登录 MySQL
root@54b354503ea5:/# mysql -uroot -p123456
mysql> use mydb;
Database changed
# 重新启动一个 mysqlsession 连接之后，临时表就不存在了
mysql> select * from tempuserinfo;
ERROR 1146 (42S02): Table 'mydb.tempuserinfo' doesn't exist
```

5.9 本章练习

1. 分别介绍 SQL 语言查询的内连接查询、左连接查询、右连接查询这三种查询各自的特点以及应该注意的事项。

2. 举例说明 in 与 exists 关键字在子查询中的应用场合。

第 6 章

MySQL 基础函数

本章内容：
- MySQL 字符串函数
- MySQL 数字函数
- MySQL 日期函数

本章将讲解 MySQL 字符串类型中常用的字符串函数、数字函数、日期函数，这些函数在数据库应用中是必不可少的。本章会列举大量示例供读者理解和学习。

6.1 MySQL 字符串函数

本节将介绍 MySQL 中字符串类型常用的函数。

6.1.1 character_length(s)：返回字符串长度

该函数用于返回字符串 s 的字符数。

例 6.1　返回字符串"clay"的字符数，示例 SQL 语句及其执行结果如下：

```
mysql> select character_length("clay") as lengthofstring;
+----------------+
| LengthOfString |
+----------------+
|              4 |
+----------------+
1 row in set (0.00 sec)
```

6.1.2　concat(s1,s2,…,sn)：字符串合并

该函数用于将 s1、s2 等多个字符串合并为一个字符串。

例 6.2　合并多个字符串，示例 SQL 语句及其执行结果如下：

```
mysql> select concat("sql", "nosql", "newsql ", "end") as concatenatedstring;
+--------------------+
| concatenatedstring |
+--------------------+
| sql nosqlnewsql end |
+--------------------+
1 row in set (0.00 sec)
```

6.1.3　format(x,n)：数字格式化

该函数可以将数字 x 格式化为 "#,###.##"，将 x 保留到小数点后 n 位，最后一位四舍五入。

例 6.3　将数字格式化为保留 2 位小数点，示例 SQL 语句及其执行结果如下：

```
mysql> select format(6500700.52341, 2);
+--------------------------+
| format(6500700.52341, 2) |
+--------------------------+
| 6,500,700.52             |
+--------------------------+
1 row in set (0.00 sec)
```

6.1.4　lpad(s1,len,s2)：字符串填充

该函数用于在字符串 s1 的开始处填充字符串 s2，使字符串长度达到 len。

例 6.4　将字符串"ABC"填充到"abc"字符串的开始处，使字符串长度为 8，示例 SQL 语句及其执行结果如下：

```
mysql> select lpad("abc",8,"ABC");
+---------------------+
| lpad("abc",8,"ABC") |
+---------------------+
| ABCABabc            |
+---------------------+
1 row in set (0.00 sec)
```

6.1.5　field(s,s1,s2,…)：返回字符串出现的位置

该函数用于返回第一个字符串 s 在字符串列表（s1,s2,…）中的位置。

例 6.5　返回字符串"c"在字符串列表中的位置，示例 SQL 语句及其执行结果如下：

```
mysql> select field("c", "a", "b", "c", "d");
+--------------------------------+
```

```
| field("c", "a", "b", "c", "d") |
+--------------------------------+
|                              3 |
+--------------------------------+
1 row in set (0.00 sec)
```

6.1.6　insert(s1,x,len,s2)：替换字符串

该函数用字符串 s2 替换字符串 s1 中从 x 位置开始长度为 len 的字符串。

例 6.6　将字符串从第一个位置开始的 4 个字符替换为"clay"，示例 SQL 语句及其执行结果如下：

```
mysql> select insert("this is aaaaaaaaa", 1, 4, "clay");
+-------------------------------------------+
| insert("this is aaaaaaaaa", 1, 4, "clay") |
+-------------------------------------------+
| clay is aaaaaaaaa                         |
+-------------------------------------------+
1 row in set (0.00 sec)
```

6.1.7　lcase(s)：把字符串中的所有字母转换为小写字母

该函数用于把字符串 s 中的所有字母转换为小写字母。

例 6.7　把字符串"CLAY"中的所有字母转换为小写字母，示例 SQL 语句及其执行结果如下：

```
mysql> select lcase("CLAY");
+---------------+
| lcase("CLAY") |
+---------------+
| clay          |
+---------------+
1 row in set (0.00 sec)
```

6.1.8　ucase(s)：把字符串中的所有字母转换为大写字母

该函数用于把字符串 s 中的所有字母转换为大写字母。

例 6.8　把字符串"clay"中的所有字母转换为大写字母，示例 SQL 语句及其执行结果如下：

```
mysql> select ucase("clay");
+---------------+
| ucase("clay") |
+---------------+
| CLAY          |
+---------------+
1 row in set (0.00 sec)
```

6.1.9　strcmp(s1,s2)：比较字符串大小

该函数用于比较字符串 s1 和 s2，如果 s1 与 s2 相等则返回 0，如果 s1>s2 则返回 1，如果 s1<s2 则返回-1。

例 6.9　比较字符串"abc"和"bcd"的大小，示例 SQL 语句及其执行结果如下：

```
mysql> select strcmp("abc", "bcd");
+----------------------+
| strcmp("abc", "bcd") |
+----------------------+
|                   -1 |
+----------------------+
1 row in set (0.00 sec)
```

6.1.10　replace(s,s1,s2)：字符串替换

该函数用字符串 s2 替换字符串 s 中的字符串 s1。

例 6.10　将字符串"abcd"中的字符串"a"替换为字符串"x"，示例 SQL 语句及其执行结果如下：

```
mysql> select replace("abcd","a","x");
+-------------------------+
| replace("abcd","a","x") |
+-------------------------+
| xbcd                    |
+-------------------------+
1 row in set (0.00 sec
```

6.1.11　position(s1 in s)：获取子字符串在字符串中出现的位置

该函数用于从字符串 s 中获取子字符串 s1 的开始位置。

例 6.11　返回字符串"abc"中子字符串"a"的位置，示例 SQL 语句及其执行结果如下：

```
mysql> select position("a" in "abc");
+------------------------+
| position("a" in "abc") |
+------------------------+
|                      1 |
+------------------------+
1 row in set (0.00 sec)
```

6.1.12　md5(s)：字符串加密

该函数用于对字符串 s 进行加密。

例 6.12　对字符串"password"进行 MD5 加密，示例 SQL 语句及其执行结果如下：

```
mysql> select md5("password");
```

```
+----------------------------------+
| md5("password")                  |
+----------------------------------+
| 319f4d26e3c536b5dd871bb2c52e3178 |
+----------------------------------+
1 row in set (0.00 sec)
```

6.1.13 inet_aton(ip)：把 IP 地址转换为数字

该函数用于把 IP 地址转换为数字。

例 6.13 把 192.168.1.201 转换为数字，示例 SQL 语句及其执行结果如下：

```
mysql> select inet_aton("192.168.1.201");
+----------------------------+
| inet_aton("192.168.1.201") |
+----------------------------+
|                 3232235977 |
+----------------------------+
1 row in set (0.00 sec)
```

6.1.14 inet_ntoa (s)：把数字转换为 IP 地址

该函数用于把数字 s 转换为 IP 地址。

例 6.14 把 3232235977 转换为 IP 地址，示例 SQL 语句及其执行结果如下：

```
mysql> select inet_ntoa(3232235977);
+-----------------------+
| inet_ntoa(3232235977) |
+-----------------------+
| 192.168.1.201         |
+-----------------------+
1 row in set (0.00 sec)
```

6.2 MySQL 数字函数

本节介绍 MySQL 中数字类型常用的函数。

6.2.1 ceil(x)：返回不小于 x 的最小整数

该函数用于返回不小于 x 的最小整数。

例 6.15 返回不小于 2.5 的最小整数，示例 SQL 语句及其执行结果如下：

```
mysql> select ceil(2.5);
+-----------+
| ceil(2.5) |
+-----------+
```

```
|          3 |
+------------+
1 row in set (0.00 sec)
```

6.2.2　ceiling(x)：返回不小于 x 的最小整数

该函数用于返回不小于 x 的最小整数，与 ceil(x)函数的作用相同。

例 6.16　返回不小于 2.5 的最小整数，示例 SQL 语句及其执行结果如下：

```
mysql> select ceiling(2.5);
+--------------+
| ceiling(2.5) |
+--------------+
|            3 |
+--------------+
1 row in set (0.00 sec)
```

6.2.3　floor(x)：返回不大于 x 的最大整数

该函数用于返回不大于 x 的最大整数。

例 6.17　返回不大于 1.5 的最大整数，示例 SQL 语句及其执行结果如下：

```
mysql> select floor(1.5);
+------------+
| floor(1.5) |
+------------+
|          1 |
+------------+
1 row in set (0.00 sec)
```

6.2.4　round(x)：返回最接近 x 的整数

该函数用于返回最接近 x 的整数。

例 6.18　返回最接近 12.23 的整数，示例 SQL 语句及其执行结果如下：

```
mysql> select round(12.23) ;
+--------------+
| round(12.23) |
+--------------+
|           12 |
+--------------+
1 row in set (0.00 sec)

mysql> select round(12.93) ;
+--------------+
| round(12.93) |
+--------------+
|           13 |
+--------------+
1 row in set (0.00 sec)
```

6.2.5　max(expression)：求最大值

该函数用于返回字段 expression 中的最大值。

例 6.19　返回数据表 userinfo 中字段 age 的最大值，示例 SQL 语句及其执行结果如下：

```
mysql> select age from userinfo;
+------+
| age  |
+------+
|  18  |
|  19  |
|  19  |
+------+
3 rows in set (0.00 sec)
mysql> select max(age) from userinfo;
+----------+
| max(age) |
+----------+
|    19    |
+----------+
1 row in set (0.00 sec)
```

6.2.6　min(expression)：求最小值

该函数用于返回字段 expression 中的最小值。

例 6.20　返回数据表 userinfo 中字段 age 的最小值，示例 SQL 语句及其执行结果如下：

```
mysql> select age from userinfo;
+------+
| age  |
+------+
|  18  |
|  19  |
|  19  |
+------+
3 rows in set (0.00 sec)
mysql> select min(age)  from userinfo;
+----------+
| min(age) |
+----------+
|    18    |
+----------+
1 row in set (0.00 sec)
```

6.2.7　sum(expression)：求总和

该函数用于返回指定字段 expression 的总和。

例 6.21　计算数据表 userinfo 中字段 age 的总和，示例 SQL 语句及其执行结果如下：

```
mysql> select age from userinfo;
```

```
+------+
| age  |
+------+
|  18  |
|  19  |
|  19  |
+------+
3 rows in set (0.00 sec)
mysql> select sum(age) from userinfo;
+----------+
| sum(age) |
+----------+
|    56    |
+----------+
1 row in set (0.00 sec)
```

6.2.8　avg(expression)：求平均值

该函数用于返回字段 expression 中的平均值。

例 6.22　返回数据表 userinfo 中 age 字段的平均值，示例 SQL 语句及其执行结果如下：

```
mysql> select age from userinfo;
+------+
| age  |
+------+
|  18  |
|  19  |
|  19  |
+------+
3 rows in set (0.00 sec)
mysql> select avg(age) from userinfo;
+----------+
| avg(age) |
+----------+
| 18.6667  |
+----------+
1 row in set (0.00 sec)
```

6.2.9　count(expression)：求总记录数

该函数用于返回查询的记录总数，expression 参数是一个字段或者"*"。

- count（字段名）：计算指定列下总的行数，计算时将忽略空值的行。
- count（*）：计算数据表中总的行数，无论某列是否为空值都包含在内。

例 6.23　返回 userinfo 表中总共有多少条记录，示例 SQL 语句及其执行结果如下：

```
mysql> select count(*) from userinfo;
+----------+
| count(*) |
+----------+
|    3     |
```

```
+----------+
1 row in set (0.00 sec)
mysql> select count(*) from userinfo where age>19;
+----------+
| count(*) |
+----------+
|        0 |
+----------+
1 row in set (0.00 sec)
```

6.3　MySQL 日期函数

本节介绍 MySQL 中日期类型常用的函数。

6.3.1　adddate(d,n)：返回指定日期加上指定天数后的日期

该函数用于计算起始日期 d 加上 n 天后的日期。

例 6.24　计算在 2021-06-06 的基础上加上 60 天后的日期，示例 SQL 语句及其执行结果如下：

```
mysql> select adddate("2017-06-15", 60);
+---------------------------+
| adddate("2017-06-15", 60) |
+---------------------------+
| 2017-08-14                |
+---------------------------+
1 row in set (0.00 sec)
```

6.3.2　addtime(t,n)：返回指定时间加上指定时间后的时间

该函数用于计算将时间 t 加上 n 秒后的时间（是一个时间表达式）。

例 6.25　计算 2021-06-06 23:23:10 分别加上 8 秒和 1 小时 10 分 5 秒后的时间，示例 SQL 语句及其执行结果如下：

```
# 2021-06-06 23:23:10 加 8 秒
mysql> select addtime("2021-06-06 23:23:10", 8);
+-----------------------------------+
| addtime("2021-06-06 23:23:10", 8) |
+-----------------------------------+
| 2021-06-06 23:23:18               |
+-----------------------------------+
1 row in set (0.00 sec)

# 2021-06-06 23:23:10 加 1 小时 10 分 5 秒
mysql> select addtime("2021-06-06 23:23:10", "1:10:5");
+------------------------------------------+
| addtime("2021-06-06 23:23:10", "1:10:5") |
```

```
+---------------------------------------+
| 2021-06-07 00:33:15                   |
+---------------------------------------+
1 row in set (0.00 sec)
```

6.3.3　curdate()：返回当前日期

该函数用于返回当前日期。

例 6.26　返回当前日期，示例 SQL 语句及其执行结果如下：

```
mysql> select curdate();
+------------+
| curdate()  |
+------------+
| 2022-06-06 |
+------------+
1 row in set (0.00 sec)
```

6.3.4　datediff(d1,d2)：返回两个日期相隔的天数

该函数用于计算日期 d1 和 d2 之间相隔的天数。

例 6.27　计算日期 2021-08-08 和 2021-06-06 之间相隔的天数，示例 SQL 语句及其执行结果如下：

```
mysql> select datediff("2021-08-08","2021-06-06");
+-------------------------------------+
| datediff("2021-08-08","2021-06-06") |
+-------------------------------------+
|                                  63 |
+-------------------------------------+
1 row in set (0.00 sec)
```

6.3.5　dayofyear(d)：返回指定日期是本年的第几天

该函数用于计算日期 d 是本年的第几天。

例 6.28　计算日期 2021-06-06 是本年的第几天，示例 SQL 语句及其执行结果如下：

```
mysql> select dayofyear('2021-06-06');
+-------------------------+
| dayofyear('2021-06-06') |
+-------------------------+
|                     157 |
+-------------------------+
1 row in set (0.00 sec)
```

6.3.6　extract(type from d)：根据对应格式返回日期

该函数用于从日期 d 中获取指定的值，type 指定返回的值。

type 可以取值为表 6-1 中的类型。

表6-1 type取值类型

取值类型	说明	取值类型	说明
hour	小时	quarter	季
minute	分	year_month	年和月
second	秒	day_hour	日和小时
microsecond	毫秒	day_minute	日和分
year	年	day_second	日和秒
month	月	hour_minute	小时和分
day	日	hour_second	小时和秒
week	周	minute_second	分和秒

例 6.29 分别获取 2021-06-06 23:11:11 日期的分钟部分、当月的天数以及当年的周数，示例 SQL 语句及其执行结果如下：

```
# 获取 2021-06-06 23:11:11 日期的分钟部分
mysql> select extract(minute from "2021-06-06 23:11:11") ;
+--------------------------------------------+
| extract(minute from '2021-06-06 23:11:11') |
+--------------------------------------------+
|                                         11 |
+--------------------------------------------+
1 row in set (0.00 sec)
# 获取 2021-06-06 23:11:11 日期所在月的天数
mysql> select extract(day from "2021-06-06 23:11:11") ;
+-----------------------------------------+
| extract(day from '2021-06-06 23:11:11') |
+-----------------------------------------+
|                                       6 |
+-----------------------------------------+
1 row in set (0.00 sec)
# 获取 2021-06-06 23:11:11 日期在当年的周数
mysql> select extract(week from '2021-06-06 23:11:11') ;
+------------------------------------------+
| extract(week from '2021-06-06 23:11:11') |
+------------------------------------------+
|                                       23 |
+------------------------------------------+
1 row in set (0.00 sec)
```

6.3.7 now()：返回当前日期和时间

该函数用于返回当前的日期和时间。

例 6.30 返回当前的日期和时间，示例 SQL 语句及其执行结果如下：

```
mysql> select now();
+---------------------+
```

```
| now()               |
+---------------------+
| 2022-06-06 16:30:23 |
+---------------------+
1 row in set (0.00 sec)
```

6.3.8　quarter(d)：返回日期对应的季度数

该函数用于返回日期 d 是第几季度，范围是 1～4。

例 6.31　返回日期 2021-06-07 是第几季度，示例 SQL 语句及其执行结果如下：

```
mysql> select quarter("2021-06-07");
+-----------------------+
| quarter("2021-06-07") |
+-----------------------+
|                     2 |
+-----------------------+
1 row in set (0.00 sec)
```

6.3.9　second(t)：返回指定时间中的秒数

该函数用于返回时间 t 中的秒数。

例 6.32　返回 2021-06-07 11:2:3 中的秒数，示例 SQL 语句及其执行结果如下：

```
mysql> select second("2021-06-07 11:2:3");
+-----------------------------+
| second("2021-06-07 11:2:3") |
+-----------------------------+
|                           3 |
+-----------------------------+
1 row in set (0.00 sec)
```

6.3.10　timediff(time1, time2)：计算时间差

该函数用于计算时间差值。

例 6.33　计算 2021-06-06 16:42:45 和 2020-06-06 16:42:45 的时间差，示例 SQL 语句及其执行结果如下：

```
mysql> select timediff("2021-06-06 16:42:45", "2020-06-06 16:42:45");
+--------------------------------------------------------+
| timediff("2021-06-06 16:42:45", "2020-06-06 16:42:45") |
+--------------------------------------------------------+
| 838:59:59                                              |
+--------------------------------------------------------+
1 row in set, 1 warning (0.00 sec)
```

6.3.11 date(t)：从指定日期时间中提取日期值

该函数用于从指定日期时间中提取日期值。

例 6.34 从 2021-06-16 23:11:11 中提取日期值，示例 SQL 语句及其执行结果如下：

```
mysql> select date("2021-06-16 23:11:11");
+-----------------------------+
| date("2021-06-16 23:11:11") |
+-----------------------------+
| 2021-06-16                  |
+-----------------------------+
1 row in set (0.00 sec)
```

6.3.12 hour(t)：返回指定时间中的小时数

该函数用于返回时间 t 中的小时数。

例 6.35 返回 2021-06-06 23:11:11 中的小时数，示例 SQL 语句及其执行结果如下：

```
mysql> select hour("2021-06-06 23:11:11");
+-----------------------------+
| hour("2021-06-06 23:11:11") |
+-----------------------------+
|                          23 |
+-----------------------------+
1 row in set (0.00 sec)
```

6.3.13 time(expression)：提取日期时间参数中的时间部分

该函数用于从传入的日期时间参数中提取时间部分。

例 6.36 提取 2021-06-06 16:42:45 的时间部分，示例 SQL 语句及其执行结果如下：

```
mysql> select time("2021-06-06 16:42:45");
+-----------------------------+
| time("2021-06-06 16:42:45") |
+-----------------------------+
| 16:42:45                    |
+-----------------------------+
1 row in set (0.00 sec)
```

6.3.14 time_format(t,f)：根据表达式显示时间

该函数用于按表达式 f 的要求显示时间 t。

例 6.37 按表达式%r 的要求显示时间 2021-06-06 16:42:45，示例 SQL 语句及其执行结果如下：

```
mysql> select time_format('2021-06-06 16:42:45','%r');
+-----------------------------------------+
```

```
| time_format('2021-06-06 16:42:45','%r') |
+-----------------------------------------+
| 04:42:45 PM                             |
+-----------------------------------------+
1 row in set (0.00 sec)
```

6.3.15　year(d)：返回指定日期的年份

该函数用于返回年份。

例 6.38　返回 2021-06-06 16:42:45 的年份，示例 SQL 语句及其执行结果如下：

```
mysql> select year("2021-06-06 16:42:45");
+-----------------------------+
| year("2021-06-06 16:42:45") |
+-----------------------------+
|                        2021 |
+-----------------------------+
1 row in set (0.00 sec)
```

第 3 部分　MySQL 高级查询函数

第 7 章　MySQL 高级函数
第 8 章　MySQL 窗口函数

第 7 章

MySQL 高级函数

本章内容：

- MySQL 表达式类型转换函数
- MySQL 返回非空表达式函数
- MySQL 表达式判断函数
- MySQL 表达式替换函数
- MySQL 返回最近生成的自增 ID 函数
- MySQL 表达式分支函数

本章将讲解 MySQL 中数据类型转换函数、返回非空表达式函数、表达式判断函数、表达式替换函数、返回最近生成的自增 ID 函数、表达式分支函数，这些函数在数据库高级引用中必不可少。本章会列举大量示例供读者理解和学习。

7.1 cast(x as type)：数据类型转换

该函数用于转换数据类型。

例 7.1 将字符串"20210607"转换为日期，示例 SQL 语句及其执行结果如下：

```
mysql> select cast("20210607" as date);
+--------------------------+
| cast("20210607" as date) |
+--------------------------+
| 2021-06-07               |
+--------------------------+
```

```
1 row in set (0.00 sec)
```

7.2　coalesce(expr1, …, exprn)：返回第一个非空表达式

该函数用于返回参数中的第一个非空表达式（从左向右）。

例 7.2　返回 null、null、"clay"、null、"clay2" 中的第一个非空表达式，示例 SQL 语句及其执行结果如下：

```
mysql> select coalesce(null, null, "clay", null, "clay2");
+---------------------------------------------+
| coalesce(null, null, "clay", null, "clay2") |
+---------------------------------------------+
| clay                                        |
+---------------------------------------------+
1 row in set (0.00 sec)
```

7.3　if(expr,v1,v2)：表达式判断

如果表达式 expr 成立，则返回结果 v1，否则返回结果 v2。

例 7.3　返回 1>0 的结果，示例 SQL 语句及其执行结果如下：

```
mysql> select if(1 > 0,'yes','no');
+----------------------+
| if(1 > 0,'yes','no') |
+----------------------+
| yes                  |
+----------------------+
1 row in set (0.00 sec)
```

7.4　ifnull(v1,v2)：null 替换

如果 v1 的值不为 null，则返回 v1；否则返回 v2。

例 7.4　返回 ifnull(null, "clay") 的结果，示例 SQL 语句及其执行结果如下：

```
mysql> select ifnull(null,"clay");
+---------------------+
| ifnull(null,"clay") |
+---------------------+
| clay                |
+---------------------+
1 row in set (0.00 sec)
```

7.5 isnull(expression)：判断表达式是否为 null

该函数用于判断表达式是否为 null，若是则返回 1，否则返回 0。

例 7.5 分别判断 isnull("clay")和 isnull(null)是否为 null，示例 SQL 语句及其执行结果如下：

```
# 判断 isnull("clay") 是否为 null
mysql> select isnull("clay");
+----------------+
| isnull("clay") |
+----------------+
|              0 |
+----------------+
1 row in set (0.00 sec)

# 判断 isnull(null)是否为 null
mysql> select isnull(null);
+--------------+
| isnull(null) |
+--------------+
|            1 |
+--------------+
1 row in set (0.00 sec)
```

7.6 nullif(expr1, expr2)：字符串相等则返回 null

该函数用于比较两个字符串，如果字符串 expr1 与 expr2 相等，则返回 null；否则返回 expr1。

例 7.6 分别比较字符串"11"和字符串"2"、字符串"a"和字符串"a"，示例 SQL 语句及其执行结果如下：

```
# 比较字符串"11"和字符串"2"
mysql> select nullif("11","2");
+------------------+
| nullif("11","2") |
+------------------+
| 11               |
+------------------+
1 row in set (0.00 sec)
# 比较字符串"a"和字符串"a"
mysql> select nullif("a","a");
+-----------------+
| nullif("a", "a") |
+-----------------+
| null            |
+-----------------+
1 row in set (0.00 sec)
```

7.7　last_insert_id()：返回最近生成的自增 ID

返回最近生成的自增 ID（auto_increment 值）。

例 7.7　返回最近生成的自增 ID，示例 SQL 语句及其执行结果如下：

```
mysql> drop table if exists userinfo;
Query OK, 0 rows affected (0.01 sec)
mysql> create table userinfo(id int(10) NOT null auto_increment primary key,
    -> name varchar(20),
    -> age int(2));
Query OK, 0 rows affected, 2 warnings (0.02 sec)
mysql> insert into userinfo (name,age) values('zhangsan',26);
Query OK, 1 row affected (0.01 sec)
mysql>  insert into userinfo (name,age) values('lisi',26);
Query OK, 1 row affected (0.01 sec)
mysql> select id from userinfo;
+----+
| id |
+----+
|  1 |
|  2 |
+----+
2 rows in set (0.00 sec)
mysql> select last_insert_id();
+------------------+
| last_insert_id() |
+------------------+
|                2 |
+------------------+
1 row in set (0.00 sec)
```

7.8　case expression：表达式分支

case expression 的语法如下：

```
case expression
    when condition1 then result1
    when condition2 then result2
    ...
    when conditionN then resultN
    else result
end
```

case 表示函数开始，end 表示函数结束。如果 condition1 成立，则返回 result1；如果 condition2 成立，则返回 result2；如果全部不成立，则返回 result。而当有一个成立之后，后面的就不执行了。

例 7.8　根据 name 分组，统计其次数，并输出其等级，示例 SQL 语句及其执行结果如下：

```
mysql> select name,(case when singin_count<5 then '<5' when singin_count>=5 and
singin_count<10 then '<10' else '>10' end ) as level
    -> from (select name, sum(singin) as singin_count from employee_tbl group by
name )t;
+------+-------+
| name | level |
+------+-------+
| clay | <10   |
| alsi | <10   |
| tom  | <5    |
+------+-------+
3 rows in set (0.00 sec)
```

7.9 本章练习

1. 举例说明 cate 与 coalesce 函数在查询中的应用场合。
2. 举例说明 if 与 ifnull 函数在查询中的应用场合。
3. 举例说明 ifnull 与 nullif 函数在查询中的应用场合。

第 8 章

MySQL 窗口函数

本章内容：
- MySQL 排名函数
- MySQL 分析函数
- MySQL 头尾函数
- MySQL 聚合统计函数

本章将讲解 MySQL 中的窗口函数，包括排名函数、分析函数、头尾函数、聚合统计函数等，这些函数在数据库数据分析和统计方面必不可少。本章会列举大量示例供读者理解和学习。

8.1 窗口函数语法和数据准备

在熟悉了 SQL 的简单查询之后，比如使用 select from where group by 这样的基础语句，若想要进一步提升自己的 SQL 技能，则一定要知道窗口函数（Window Function），它又被称为分析函数（Analytics Function）。窗口函数类似于可以返回聚合值的函数，例如 sum()、count()、max()。但是窗口函数又与普通的聚合函数不同，它不会对结果进行分组，因此输出中的行数与输入中的行数相同。窗口函数语法如下：

```
SELECT SUM() OVER(PARTITION BY ___ ORDER BY___) FROM Table;
```

参数说明：

- over 关键字用来指定函数执行的窗口范围，如果后面括号中什么都不写，则意味着窗口包含满足 where 条件的所有数据行，窗口函数基于所有数据行进行计算；如果括号中不为空，则支持以下两种语法来设置窗口：
- ➤ partition by 子句：窗口按照哪些字段进行分组，窗口函数在不同的分组上分别执行。
- ➤ order by 子句：窗口按照哪些字段进行排序，窗口函数将按照排序后的记录顺序进行编号。

窗口函数包括排名函数、分析函数、头尾函数、聚合统计函数和其他函数。

- 排名函数：用于对查询结果中的每一行进行排名，常见的排名函数有 row_number()、rank()、dense_rank()等。
- 分析函数：用于在查询结果中对数据进行分析。常见的分析函数有 lag()、lead()等。
- 头尾函数：用于获取分组数据中每个分组的第一个或最后一个元素，常见的头尾函数有 first_value()和 last_value()。
- 聚合统计函数：用于对查询结果进行聚合统计，常见的聚合统计函数有 count()、sum()、avg()、max()和 min()等。
- 其他函数：常用的其他窗口函数还有 nth_value()、ntile()函数。

接下来我们通过一张学生成绩表来学习这些常用的窗口函数。

```sql
# 数据准备，创建一张学生成绩表
create table student_score (
    stu_no int(10),   --学生编号
    course_no varchar(50),--科目编号
    score  int   --科目成绩
);
# 插入测试数据
insert into student_score (stu_no ,course_no,score) values (1,'L0001',90);
insert into student_score (stu_no ,course_no,score) values (1,'L0002',98);
insert into student_score (stu_no ,course_no,score) values (1,'L0003',98);
insert into student_score (stu_no ,course_no,score) values (1,'L0004',95);
insert into student_score (stu_no ,course_no,score) values (1,'L0005',95);
insert into student_score (stu_no ,course_no,score) values (2,'L0001',90);
insert into student_score (stu_no ,course_no,score) values (2,'L0002',98);
insert into student_score (stu_no ,course_no,score) values (2,'L0003',98);
insert into student_score (stu_no ,course_no,score) values (2,'L0004',95);
insert into student_score (stu_no ,course_no,score) values (2,'L0005',96);
insert into student_score (stu_no ,course_no,score) values (3,'L0001',98);
insert into student_score (stu_no ,course_no,score) values (3,'L0002',96);
insert into student_score (stu_no ,course_no,score) values (3,'L0003',90);
insert into student_score (stu_no ,course_no,score) values (3,'L0004',98);
insert into student_score (stu_no ,course_no,score) values (3,'L0005',97);
# 查看数据
mysql> select * from student_score;
+--------+-----------+-------+
| stu_no | course_no | score |
+--------+-----------+-------+
|      1 | L0001     |    90 |
|      1 | L0002     |    98 |
|      1 | L0003     |    98 |
|      1 | L0004     |    95 |
|      2 | L0001     |    90 |
|      2 | L0002     |    98 |
```

```
|   2    | L0003      |  98  |
|   2    | L0004      |  95  |
|   3    | L0001      |  98  |
|   3    | L0002      |  96  |
|   3    | L0003      |  90  |
|   3    | L0004      |  98  |
|   1    | L0005      |  95  |
|   2    | L0005      |  96  |
|   3    | L0005      |  97  |
+--------+------------+------+
15 rows in set (0.00 sec)
```

8.2 排名函数

8.2.1 row_number()：顺序排名函数

该函数根据提供的字段的值依次进行排序，如果出现相同的值，排名号仍会依次增加。比如，对（10，20，20）这个集合排名，排名序号依次是 1、2、3。

例 8.1 对学生编号是 1 的学生的各科目成绩进行排名，示例 SQL 语句及其执行结果如下：

```
mysql> select *
    -> from (
    -> select stu_no, row_number() over (partition by stu_no
    -> order by score desc ) as score_order, course_no, score
    -> from student_score
    -> ) t
    -> where stu_no =1;
+--------+-------------+-----------+-------+
| stu_no | score_order | course_no | score |
+--------+-------------+-----------+-------+
|   1    |      1      | L0002     |  98   |
|   1    |      2      | L0003     |  98   |
|   1    |      3      | L0004     |  95   |
|   1    |      4      | L0005     |  95   |
|   1    |      5      | L0001     |  90   |
+--------+-------------+-----------+-------+
5 rows in set (0.00 sec)
```

如上所示，尽管该学生的科目 L0002 和 L0003 的成绩一样，科目 L0004 和 L0005 的成绩一样，但是排名还是依次增加的。

例 8.2 查询每个学生的分数最高的科目，示例 SQL 语句及其执行结果如下：

```
mysql> select *
    -> from (
    -> select stu_no, row_number() over (partition by stu_no
    -> order by score desc ) as score_order, course_no, score
    -> from student_score
```

```
    -> ) t
    -> where score_order <=1 ;
+--------+-------------+------------+-------+
| stu_no | score_order | course_no  | score |
+--------+-------------+------------+-------+
|      1 |           1 | L0002      |    98 |
|      2 |           1 | L0002      |    98 |
|      3 |           1 | L0001      |    98 |
+--------+-------------+------------+-------+
3 rows in set (0.00 sec)
```

如上所示，编号为 1、2、3 的学生的分数最高的科目分别是 L0002、L0002、L0001，结果符合原始数据的答案。

8.2.2 rank()：跳级排名函数

该函数将根据提供的字段的值进行排序（跳过重复序号），当出现相同的值时，排名号会一致，不过下一个排名号是上一个排名号加上重复的排名号。比如，对（10, 20, 20, 30, 30）这个集合排名，排名序号依次是 1、1、3、4、4。

例 8.3　对学生编号是 1 的学生的各科目成绩进行排名，示例 SQL 语句及其执行结果如下：

```
mysql> select *
    -> from (
    -> select stu_no, rank() over (partition by stu_no
    -> order by score desc ) as score_order, course_no, score
    -> from student_score
    -> ) t
    -> where stu_no =1 ;
+--------+-------------+------------+-------+
| stu_no | score_order | course_no  | score |
+--------+-------------+------------+-------+
|      1 |           1 | L0002      |    98 |
|      1 |           1 | L0003      |    98 |
|      1 |           3 | L0004      |    95 |
|      1 |           3 | L0005      |    95 |
|      1 |           5 | L0001      |    90 |
+--------+-------------+------------+-------+
5 rows in set (0.00 sec)
```

如上所示，编号是 1 的学生的科目 L0002 和 L0003 的成绩一致，所以排名都是 1；然后下一个成绩的排名直接跳过 2 排名为 3，因为科目 L0004 和 L0005 的成绩一致，所以排名都是 3；接着下一个成绩的排名直接跳过 4 排名为 5。

例 8.4　查询每个学生的分数最高的科目，示例 SQL 语句及其执行结果如下：

```
mysql> select *
    -> from (
    -> select stu_no, rank() over (partition by stu_no
    -> order by score desc) as score_order, course_no, score
```

```
    -> from student_score
    -> ) t
    -> where score_order <= 1;
+--------+-------------+-----------+-------+
| stu_no | score_order | course_no | score |
+--------+-------------+-----------+-------+
|      1 |           1 | L0002     |    98 |
|      1 |           1 | L0003     |    98 |
|      2 |           1 | L0002     |    98 |
|      2 |           1 | L0003     |    98 |
|      3 |           1 | L0001     |    98 |
|      3 |           1 | L0004     |    98 |
+--------+-------------+-----------+-------+
6 rows in set (0.00 sec)
```

如上所示，一名学生的分数最高的科目存在多个，这是因为使用 rank() 排名导致出现重复的排名。在实际业务中需要依据实际情况选择合适的排名函数。

8.2.3 dense_rank()：不跳级排名函数

该函数将根据提供的字段的值进行排序（不跳过序号），当出现相同的值时，排名号会一致，不过下一个排名号是上一个排名号加 1。比如，对（10, 20, 20, 30, 30）这个集合排名，排名序号依次是 1、1、2、3、3。

例 8.5 对学生编号是 2 的学生的各科目成绩进行排名，示例 SQL 语句及其执行结果如下：

```
mysql> select *
    -> from (
    -> select stu_no, dense_rank() over (partition by stu_no
    -> order by score desc ) as score_order, course_no, score
    -> from student_score
    -> ) t
    -> where stu_no =2 ;
+--------+-------------+-----------+-------+
| stu_no | score_order | course_no | score |
+--------+-------------+-----------+-------+
|      2 |           1 | L0002     |    98 |
|      2 |           1 | L0003     |    98 |
|      2 |           2 | L0005     |    96 |
|      2 |           3 | L0004     |    95 |
|      2 |           4 | L0001     |    90 |
+--------+-------------+-----------+-------+
5 rows in set (0.00 sec)
```

如上所示，编号为 2 的学生的科目 L0002 和 L0003 的成绩一致，所以排名都是 1，然后下一个成绩的科目排名顺序为 2。

8.3 分析函数

8.3.1 lag()：前分析函数，返回前 n 行的值

lag(expr,n)函数用于返回当前行的前 n 行的 expr 的值。内层 SQL 先通过 lag()函数得到前 n 个值，外层 SQL 再将当前值与前 n 个值相减得到差值 diff。

例 8.6 查询科目 L0001 的成绩，根据分数排名，查看后一名和前一名的成绩之差，示例 SQL 语句及其执行结果如下：

```
mysql> select stu_no, course_no, score, pre_score,
    -> score-pre_score as diff
    -> from(
    ->     select stu_no, course_no, score,
    ->     lag(score,1) over w as pre_score
    ->     from student_score
    ->     where course_no in ('l0001')
    ->     window w as (partition by course_no order by score)) t
    -> ;
+--------+-----------+-------+-----------+------+
| stu_no | course_no | score | pre_score | diff |
+--------+-----------+-------+-----------+------+
|      1 | L0001     |    90 |      null | null |
|      2 | L0001     |    90 |        90 |    0 |
|      3 | L0001     |    98 |        90 |    8 |
+--------+-----------+-------+-----------+------+
3 rows in set (0.00 sec)
```

如上所示，最后一行 diff 代表的是后一名学生的成绩和上一名学生的成绩之差。

例 8.7 查询所有科目的成绩，根据分数排名查看前面两名学生的成绩和当前这名学生的成绩之差，示例 SQL 语句及其执行结果如下：

```
mysql> select stu_no, course_no, score, pre_score,
    -> score-pre_score as diff
    -> from(
    ->     select stu_no, course_no, score,
    ->     lag(score,2) over w as pre_score
    ->     from student_score
    ->     window w as (partition by course_no order by score)) t
    -> ;
+--------+-----------+-------+-----------+------+
| stu_no | course_no | score | pre_score | diff |
+--------+-----------+-------+-----------+------+
|      1 | L0001     |    90 |      null | null |
|      2 | L0001     |    90 |      null | null |
|      3 | L0001     |    98 |        90 |    8 |
|      3 | L0002     |    96 |      null | null |
|      1 | L0002     |    98 |      null | null |
|      2 | L0002     |    98 |        96 |    2 |
|      3 | L0003     |    90 |      null | null |
```

```
|    1   | L0003     |   98  |   null     | null |
|    2   | L0003     |   98  |    90      |   8  |
|    1   | L0004     |   95  |   null     | null |
|    2   | L0004     |   95  |   null     | null |
|    3   | L0004     |   98  |    95      |   3  |
|    1   | L0005     |   95  |   null     | null |
|    2   | L0005     |   96  |   null     | null |
|    3   | L0005     |   97  |    95      |   2  |
+--------+-----------+-------+------------+------+
15 rows in set (0.00 sec)
```

8.3.2 lead()：后分析函数，返回后 n 行的值

lead(expr,n) 函数用于返回当前行的后 n 行的 expr 的值。内层 SQL 先通过 lead() 函数得到后 n 个值，外层 SQL 再将当前值与后 n 个值相减得到差值 diff。

例 8.8 查询科目 L0001 的成绩，根据分数排名，查看前一名和后一名的成绩之差，示例 SQL 语句及其执行结果如下：

```
mysql> select stu_no, course_no, score, pre_score,
    -> score-pre_score as diff
    -> from(
    ->    select stu_no, course_no, score,
    ->    lead(score,1) over w as pre_score
    ->    from student_score
    ->    where course_no in ('l0001')
    ->    window w as (partition by course_no order by score)) t;
+--------+-----------+-------+------------+------+
| stu_no | course_no | score | pre_score  | diff |
+--------+-----------+-------+------------+------+
|    1   | L0001     |   90  |    90      |   0  |
|    2   | L0001     |   90  |    98      |  -8  |
|    3   | L0001     |   98  |   null     | null |
+--------+-----------+-------+------------+------+
3 rows in set (0.00 sec)
```

如上所示，结果符合预期。

8.4 头 尾 函 数

8.4.1 first_value()：头尾函数，返回第一个值

first_value(expr) 函数用于返回第一行的 expr 的值。

例 8.9 根据科目分组对成绩排序，查询科目 L0005 的成绩中第一个（最低分）成绩值，示例 SQL 语句及其执行结果如下：

```
mysql> select stu_no, course_no, score,
    -> first_value(score) over (partition by course_no order by score) as
```

```
first_score
    -> from student_score
    -> where course_no in ('l0005')
    -> ;
+--------+-----------+-------+-------------+
| stu_no | course_no | score | first_score |
+--------+-----------+-------+-------------+
|      1 | L0005     |    95 |          95 |
|      2 | L0005     |    96 |          95 |
|      3 | L0005     |    97 |          95 |
+--------+-----------+-------+-------------+
3 rows in set (0.01 sec)
```

如上所示，最后的字段 first_score 表示科目成绩排名后的第一个成绩值，结果符合预期。

例 8.10 根据科目分组对成绩排序，查询每科的成绩中第一个（最低分）成绩值，示例 SQL 语句及其执行结果如下：

```
mysql> select stu_no, course_no, score,
    -> first_value(score) over (partition by course_no order by score) as first_score
    -> from student_score
    -> ;
+--------+-----------+-------+-------------+
| stu_no | course_no | score | first_score |
+--------+-----------+-------+-------------+
|      1 | L0001     |    90 |          90 |
|      2 | L0001     |    90 |          90 |
|      3 | L0001     |    98 |          90 |
|      3 | L0002     |    96 |          96 |
|      1 | L0002     |    98 |          96 |
|      2 | L0002     |    98 |          96 |
|      3 | L0003     |    90 |          90 |
|      1 | L0003     |    98 |          90 |
|      2 | L0003     |    98 |          90 |
|      1 | L0004     |    95 |          95 |
|      2 | L0004     |    95 |          95 |
|      3 | L0004     |    98 |          95 |
|      1 | L0005     |    95 |          95 |
|      2 | L0005     |    96 |          95 |
|      3 | L0005     |    97 |          95 |
+--------+-----------+-------+-------------+
15 rows in set (0.00 sec)
```

如上所示，结果符合预期。

8.4.2 last_value()：头尾函数，返回最后一个值

last_value(expr)函数返回最后一行的 expr 的值。

例 8.11 根据科目分组对成绩排序，查询科目 L0005 的成绩中最后一个（最高分）成绩值，示例 SQL 语句及其执行结果如下：

```
mysql> select stu_no, course_no, score,
    -> last_value(score) over (partition by course_no order by score range between unbounded preceding and unbounded following) as last_score
    -> from student_score
    -> where course_no in ('l0005')
    -> ;
+--------+-----------+-------+------------+
| stu_no | course_no | score | last_score |
+--------+-----------+-------+------------+
|      1 | L0005     |    95 |         97 |
|      2 | L0005     |    96 |         97 |
|      3 | L0005     |    97 |         97 |
+--------+-----------+-------+------------+
3 rows in set (0.00 sec)
```

如上结果所示，字段 last_score 表示科目成绩排名后的最后一个成绩值，结果符合预期。

例 8.12 根据科目分组对成绩排序，查询每个科目的成绩中最后一个（最高分）成绩值，示例 SQL 语句及其执行结果如下：

```
mysql> select stu_no, course_no, score,
    -> last_value(score) over (partition by course_no order by score range between unbounded preceding and unbounded following) as last_score
    -> from student_score
    -> ;
+--------+-----------+-------+------------+
| stu_no | course_no | score | last_score |
+--------+-----------+-------+------------+
|      1 | L0001     |    90 |         98 |
|      2 | L0001     |    90 |         98 |
|      3 | L0001     |    98 |         98 |
|      3 | L0002     |    96 |         98 |
|      1 | L0002     |    98 |         98 |
|      2 | L0002     |    98 |         98 |
|      3 | L0003     |    90 |         98 |
|      1 | L0003     |    98 |         98 |
|      2 | L0003     |    98 |         98 |
|      1 | L0004     |    95 |         98 |
|      2 | L0004     |    95 |         98 |
|      3 | L0004     |    98 |         98 |
|      1 | L0005     |    95 |         97 |
|      2 | L0005     |    96 |         97 |
|      3 | L0005     |    97 |         97 |
+--------+-----------+-------+------------+
15 rows in set (0.00 sec)
```

以上结果符合预期。

例 8.13 根据科目分组对成绩排序，查询每个科目的成绩中最高分和最低分的成绩值，示例 SQL 语句及其执行结果如下：

```
mysql> select stu_no, course_no, score,
    -> first_value(score) over (partition by course_no order by score) as first_score,
    -> last_value(score) over (partition by course_no order by score range between
```

```
unbounded preceding and unbounded following) as last_score
    -> from student_score
    -> ;
+--------+-----------+-------+-------------+------------+
| stu_no | course_no | score | first_score | last_score |
+--------+-----------+-------+-------------+------------+
|      1 | L0001     |    90 |          90 |         98 |
|      2 | L0001     |    90 |          90 |         98 |
|      3 | L0001     |    98 |          90 |         98 |
|      3 | L0002     |    96 |          96 |         98 |
|      1 | L0002     |    98 |          96 |         98 |
|      2 | L0002     |    98 |          96 |         98 |
|      3 | L0003     |    90 |          90 |         98 |
|      1 | L0003     |    98 |          90 |         98 |
|      2 | L0003     |    98 |          90 |         98 |
|      1 | L0004     |    95 |          95 |         98 |
|      2 | L0004     |    95 |          95 |         98 |
|      3 | L0004     |    98 |          95 |         98 |
|      1 | L0005     |    95 |          95 |         97 |
|      2 | L0005     |    96 |          95 |         97 |
|      3 | L0005     |    97 |          95 |         97 |
+--------+-----------+-------+-------------+------------+
15 rows in set (0.00 sec)
```

如上所示，first_score 表示每科最低分的成绩，last_score 表示每科最高分的成绩，结果符合预期。

8.5 聚合统计函数

8.5.1 sum()：聚合统计总和

在窗口中每条记录动态地应用聚合函数 sum()，可以动态计算在指定窗口内的 sum 函数值。

例 8.14 查询学生编号是 2 的学生的累计分数，SQL 语句及其执行结果如下：

```
mysql> select stu_no, course_no, score,
    -> sum(score) over (partition by stu_no order by course_no) as score_sum
    -> from student_score
    -> where stu_no =2
    -> ;
+--------+-----------+-------+-----------+
| stu_no | course_no | score | score_sum |
+--------+-----------+-------+-----------+
|      2 | L0001     |    90 |        90 |
|      2 | L0002     |    98 |       188 |
|      2 | L0003     |    98 |       286 |
|      2 | L0004     |    95 |       381 |
|      2 | L0005     |    96 |       477 |
+--------+-----------+-------+-----------+
5 rows in set (0.00 sec)
```

如上所示，score_sum 字段表示累计的分数，结果符合预期。

8.5.2　avg()：聚合统计平均值

在窗口中每条记录动态地应用聚合函数 avg()，可以动态计算在指定窗口内的 avg 函数值。

例 8.15　查询学生编号为 2 的学生的平均分数，示例 SQL 语句及其执行结果如下：

```
mysql> select stu_no, course_no, score,
    -> avg(score) over (partition by stu_no order by course_no  range between unbounded preceding and unbounded following) as score_avg
    -> from student_score
    -> where stu_no =2;
+--------+-----------+-------+-----------+
| stu_no | course_no | score | score_avg |
+--------+-----------+-------+-----------+
|      2 | L0001     |    90 |   95.4000 |
|      2 | L0002     |    98 |   95.4000 |
|      2 | L0003     |    98 |   95.4000 |
|      2 | L0004     |    95 |   95.4000 |
|      2 | L0005     |    96 |   95.4000 |
+--------+-----------+-------+-----------+
5 rows in set (0.00 sec)
```

如上所示，score_avg 字段表示该学生的平均分数，结果符合预期。

8.5.3　max()：聚合统计最大值

在窗口中每条记录动态地应用聚合函数 max()，可以动态计算在指定窗口内的 max 函数值。

例 8.16　查询学生编号为 2 的学生的最高分数，示例 SQL 语句及其执行结果如下：

```
mysql> select stu_no, course_no, score,
    -> max(score) over (partition by stu_no order by course_no  range between unbounded preceding and unbounded following) as score_max
    -> from student_score
    -> where stu_no =2;
+--------+-----------+-------+-----------+
| stu_no | course_no | score | score_max |
+--------+-----------+-------+-----------+
|      2 | L0001     |    90 |        98 |
|      2 | L0002     |    98 |        98 |
|      2 | L0003     |    98 |        98 |
|      2 | L0004     |    95 |        98 |
|      2 | L0005     |    96 |        98 |
+--------+-----------+-------+-----------+
5 rows in set (0.00 sec)
```

如上所示，score_max 字段表示该学生的最高分数，结果符合预期。

8.5.4 min()：聚合统计最小值

在窗口中每条记录动态地应用聚合函数 min()，可以动态计算在指定窗口内的 min 函数值。

例 8.17 根据学生编号分组并对科目编号排序，查询所有学生的最低分数，示例 SQL 语句及其执行结果如下：

```
mysql> select stu_no, course_no, score,
    -> min(score) over (partition by stu_no order by course_no  ) as score_min
    -> from student_score ;
+--------+-----------+-------+-----------+
| stu_no | course_no | score | score_min |
+--------+-----------+-------+-----------+
|      1 | L0001     |    90 |        90 |
|      1 | L0002     |    98 |        90 |
|      1 | L0003     |    98 |        90 |
|      1 | L0004     |    95 |        90 |
|      1 | L0005     |    95 |        90 |
|      2 | L0001     |    90 |        90 |
|      2 | L0002     |    98 |        90 |
|      2 | L0003     |    98 |        90 |
|      2 | L0004     |    95 |        90 |
|      2 | L0005     |    96 |        90 |
|      3 | L0001     |    98 |        98 |
|      3 | L0002     |    96 |        96 |
|      3 | L0003     |    90 |        90 |
|      3 | L0004     |    98 |        90 |
|      3 | L0005     |    97 |        90 |
+--------+-----------+-------+-----------+
15 rows in set (0.00 sec)
```

如上所示，score_min 字段表示该学生的最低分数，结果符合预期。

8.5.5 count()：聚合统计总条数

在窗口中每条记录动态地应用聚合函数 count()，可以动态计算在指定窗口内的 count 函数值。

例 8.18 计算每个学生参加考试的总科目数量，示例 SQL 语句及其执行结果如下：

```
mysql> select stu_no, course_no, score,
    -> count(score) over (partition by stu_no order by course_no  range between unbounded preceding and unbounded following ) as score_count
    -> from student_score ;
+--------+-----------+-------+-------------+
| stu_no | course_no | score | score_count |
+--------+-----------+-------+-------------+
|      1 | L0001     |    90 |           5 |
|      1 | L0002     |    98 |           5 |
|      1 | L0003     |    98 |           5 |
|      1 | L0004     |    95 |           5 |
|      1 | L0005     |    95 |           5 |
|      2 | L0001     |    90 |           5 |
|      2 | L0002     |    98 |           5 |
```

```
|    2   |  L0003    |   98  |      5       |
|    2   |  L0004    |   95  |      5       |
|    2   |  L0005    |   96  |      5       |
|    3   |  L0001    |   98  |      5       |
|    3   |  L0002    |   96  |      5       |
|    3   |  L0003    |   90  |      5       |
|    3   |  L0004    |   98  |      5       |
|    3   |  L0005    |   97  |      5       |
+--------+-----------+-------+--------------+
15 rows in set (0.00 sec)
```

如上所示，score_count 字段表示每个学生考试的总科目数量，结果符合预期。

例 8.19 根据学生编号分组来查询每个学生科目的平均分、累加分和最高分数，示例 SQL 语句及其执行结果如下：

```
mysql> select stu_no, course_no, score
    -> , sum(score) over (partition by stu_no order by course_no) as score_sum
    -> , avg(score) over (partition by stu_no order by course_no range between
unbounded preceding and unbounded following) as score_avg
    -> , max(score) over (partition by stu_no order by course_no range between
unbounded preceding and unbounded following) as score_max
    -> from student_score;
+--------+-----------+-------+-----------+-----------+-----------+
| stu_no | course_no | score | score_sum | score_avg | score_max |
+--------+-----------+-------+-----------+-----------+-----------+
|    1   |  L0001    |   90  |     90    |  95.2000  |     98    |
|    1   |  L0002    |   98  |    188    |  95.2000  |     98    |
|    1   |  L0003    |   98  |    286    |  95.2000  |     98    |
|    1   |  L0004    |   95  |    381    |  95.2000  |     98    |
|    1   |  L0005    |   95  |    476    |  95.2000  |     98    |
|    2   |  L0001    |   90  |     90    |  95.4000  |     98    |
|    2   |  L0002    |   98  |    188    |  95.4000  |     98    |
|    2   |  L0003    |   98  |    286    |  95.4000  |     98    |
|    2   |  L0004    |   95  |    381    |  95.4000  |     98    |
|    2   |  L0005    |   96  |    477    |  95.4000  |     98    |
|    3   |  L0001    |   98  |     98    |  95.8000  |     98    |
|    3   |  L0002    |   96  |    194    |  95.8000  |     98    |
|    3   |  L0003    |   90  |    284    |  95.8000  |     98    |
|    3   |  L0004    |   98  |    382    |  95.8000  |     98    |
|    3   |  L0005    |   97  |    479    |  95.8000  |     98    |
+--------+-----------+-------+-----------+-----------+-----------+
15 rows in set (0.00 sec)
```

如上所示，结合原始数据，结果符合预期。

8.6 其他函数

8.6.1 nth_value()：从结果集中的第 n 行获取值

nth_value(expr,n) 函数用于返回窗口中第 n 个 expr 的值。

例 8.20　显示编号为 1 和 2 的学生的成绩中排名第 1 和第 2 的成绩的分数，示例 SQL 语句及其执行结果如下：

```
mysql> select stu_no, course_no, score,
    -> nth_value(score,1) over w as score1,
    -> nth_value(score,2) over w as score2
    -> from student_score
    -> where stu_no in (1,2)
    -> window w as (partition by stu_no order by score desc range between unbounded preceding and unbounded following)
    -> ;
+--------+-----------+-------+--------+--------+
| stu_no | course_no | score | score1 | score2 |
+--------+-----------+-------+--------+--------+
|      1 | L0002     |    98 |     98 |     98 |
|      1 | L0003     |    98 |     98 |     98 |
|      1 | L0004     |    95 |     98 |     98 |
|      1 | L0005     |    95 |     98 |     98 |
|      1 | L0001     |    90 |     98 |     98 |
|      2 | L0002     |    98 |     98 |     98 |
|      2 | L0003     |    98 |     98 |     98 |
|      2 | L0005     |    96 |     98 |     98 |
|      2 | L0004     |    95 |     98 |     98 |
|      2 | L0001     |    90 |     98 |     98 |
+--------+-----------+-------+--------+--------+
10 rows in set (0.00 sec)
```

如上所示，score1 是第 1 名的成绩，score2 是第 2 名的成绩，结果符合预期。

例 8.21　显示所有学生的成绩中排名第 2 和第 3 的成绩的分数，示例 SQL 语句及其执行结果如下：

```
mysql> select stu_no, course_no, score,
    -> nth_value(score,2) over w as score2,
    -> nth_value(score,3) over w as score3
    -> from student_score
    -> window w as (partition by stu_no order by score desc range between unbounded preceding and unbounded following)
    -> ;
+--------+-----------+-------+--------+--------+
| stu_no | course_no | score | score2 | score3 |
+--------+-----------+-------+--------+--------+
|      1 | L0002     |    98 |     98 |     95 |
|      1 | L0003     |    98 |     98 |     95 |
|      1 | L0004     |    95 |     98 |     95 |
|      1 | L0005     |    95 |     98 |     95 |
|      1 | L0001     |    90 |     98 |     95 |
|      2 | L0002     |    98 |     98 |     96 |
|      2 | L0003     |    98 |     98 |     96 |
|      2 | L0005     |    96 |     98 |     96 |
|      2 | L0004     |    95 |     98 |     96 |
|      2 | L0001     |    90 |     98 |     96 |
|      3 | L0001     |    98 |     98 |     97 |
```

```
|      3 | L0004      |     98 |      98 |     97 |
|      3 | L0005      |     97 |      98 |     97 |
|      3 | L0002      |     96 |      98 |     97 |
|      3 | L0003      |     90 |      98 |     97 |
+--------+------------+--------+---------+--------+
15 rows in set (0.00 sec)
```

如上所示，score2 是第 2 名的成绩，score3 是第 3 名的成绩，结果符合预期。

8.6.2　ntile()：数据集分桶

ntile(n)函数将分区中的有序数据分为 n 个等级，记录等级数。此函数在数据分析中应用较多，比如由于数据量大，需要将数据平均分配到 n 个并行的进程中分别进行计算，此时就可以用 ntile(n)对数据进行分组（由于记录数不一定被 n 整除，因此数据不一定完全平均分配），然后将不同桶号的数据再进行分配，最后计算。

例 8.22　根据科目目录将数据分成两组，示例 SQL 语句及其执行结果如下：

```
mysql> select
    -> ntile(2) over w as nf,
    -> stu_no, course_no, score
    -> from student_score
    -> window w as (partition by course_no order by score)
    -> ;
+----+--------+-----------+-------+
| nf | stu_no | course_no | score |
+----+--------+-----------+-------+
|  1 |      1 | L0001     |    90 |
|  1 |      2 | L0001     |    90 |
|  2 |      3 | L0001     |    98 |
|  1 |      3 | L0002     |    96 |
|  1 |      1 | L0002     |    98 |
|  2 |      2 | L0002     |    98 |
|  1 |      3 | L0003     |    90 |
|  1 |      1 | L0003     |    98 |
|  2 |      2 | L0003     |    98 |
|  1 |      1 | L0004     |    95 |
|  1 |      2 | L0004     |    95 |
|  2 |      3 | L0004     |    98 |
|  1 |      1 | L0005     |    95 |
|  1 |      2 | L0005     |    96 |
|  2 |      3 | L0005     |    97 |
+----+--------+-----------+-------+
15 rows in set (0.00 sec)
```

如上所示，nf 代表分组的编码，结果符合预期。

8.7 本章练习

1. 举例说明顺序排名函数与跳级排名函数在查询中的应用场合以及二者的主要区别。
2. 举例说明跳级排名函数与不跳级排名函数在查询中应用场合以及二者的主要区别。
3. 举例说明头尾函数与分析函数在查询中的应用场合。
4. 分别说出 sum()、avg()、max()、min()和 count()函数表示的意义,并列出各函数的使用场景。

第 4 部分　MySQL 数据表分区

第 9 章　MySQL 表分区
第 10 章　MySQL 分区管理

第 9 章

MySQL 表分区

本章内容：

- MySQL 分区类型
- MySQL 范围分区
- MySQL 列表分区
- MySQL 列分区
- MySQL 哈希分区
- MySQL 键分区
- MySQL 子分区

本章将讲解 MySQL 的各种分区策略，如范围分区、列表分区、列分区、哈希分区、键分区、子分区等。本章会列举大量示例供读者理解和学习。

9.1 分区介绍

MySQL 目前仅支持使用 InnoDB 和 NDB 存储引擎对数据表进行分区，不支持其他存储引擎。如果使用 MyISAM 存储引擎创建数据表的分区，将会显示 ER_CHECK_NOT_IMPLEMENTED 的错误。

MySQL 分区让用户可以根据自己设置的规则把各个数据表中的数据分布在不同的文件系统中。实际上，数据表的相关部分作为单独的表存储在不同的位置上。用户设置数据划分的规则称为分区函数，在 MySQL 中分区函数可以是模数、对一组范围或值列表的简单匹配、内部哈希函数（或称为散列函数）或线性哈希函数。该函数根据用户指定的分区类型进行选择，并将用户提供的表达式的值作为其参数。此表达式可以是列值，也可以是作用于一个或多个列值的函数，具体取决于所使用的分区类型。

使用分区的优点如下：

- 数据表被分区后，其中的数据可以分布在不同的物理设备上，从而高效地利用多个硬件设备。
- 分区上的数据更容易维护。例如，想批量删除大量数据时，可以使用清除整个分区的方式来处理。另外，还可以对一个独立分区进行优化、检查、修复等操作。
- 可以使用分区来避免某些特殊的瓶颈，例如 InnoDB 的单个索引的互斥访问。
- 在大数据集的应用场景下，可以备份和恢复独立的分区，这样能够更好地提高性能。
- 某些查询也可以被极大地优化，因为满足给定 where 子句的数据只能存储在一个或多个分区上，所以会自动搜索相关分区数据，而不是扫描所有的表数据。
- 由于在创建分区后可以更改分区，因此用户可以重新组织数据，提高查询效率。

9.2 分区类型

MySQL 目前支持多种分区，比如：范围分区、列表分区、列分区、哈希分区、键分区、子分区。MySQL 的所有分区类型说明如下：

- 范围（range）分区：基于一个给定连续区间的列值，把区间列值对应的多行分配给分区。
- 列表（list）分区：类似于按范围进行分区，但不同之处在于列表分区是根据列值与离散值集合中的某个值匹配来选择的。
- 列（column）分区：数据根据某个或多个列的值进行划分，是列表分区和范围分区的变体。
- 哈希（hash）分区：基于用户定义的表达式的返回值进行分区选择，该表达式使用将要插入表中的行的列值来进行计算。哈希函数可以包含在 MySQL 中有效且产生非负整数值的表达式。
- 键（key）分区：类似于按哈希分区，区别在于键分区只支持计算一列或多列，并且 MySQL 服务器为此提供了自身的哈希函数。
- 子分区：又被称为复合分区，是对分区表中每个分区的进一步划分。

9.3 范围分区

数据表按范围分区的方式是，每个分区都包含其分区表达式的列值在给定范围内的行。范围应该是连续且不重叠的，使用 value less than 运算符来定义。

例 9.1 创建一个保存商店的员工薪资的信息表，然后对表进行分区。

首先创建表，示例 SQL 语句如下：

```
create table employees
(
    empno varchar(20) not null,
    empname varchar(20),
    deptno int,
    birthdate date,
    salary int
);
```

然后对表进行分区，该表可以通过多种方式按范围进行分区。

（1）使用 salary 列，通过添加 partition by range 子句将表分为 4 个分区：

```
create table employees
(
    empno varchar(20) not null,
    empname varchar(20),
    deptno int,
    birthdate date,
    salary int
)
partition by range(salary)(
    partition p0 values less than (5000),
    partition p1 values less than (10000),
    partition p2 values less than (15000),
    partition p3 values less than (20000)
);
```

在此分区方案中，工资在 0～5000 的员工所对应的数据行都存储在 partition 中的 p0 分区中，工资在 5000～10000 的员工所对应的数据行都存储在 partition 中的 p1 分区中，以此类推。每个分区都是按顺序从最低到最高定义的。

如果在数据表中添加工资是 15800 的员工信息：

```
(1,'zhangsan1',1,'2021-06-20',15800)
```

那么这条信息会存储在 p3 分区中。不过，当我们需要添加工资在 20000 以上的员工的信息时，在这种方案下，由于没有规则覆盖 salary 大于 20000 的行，因此会出错，因为服务器不知道该将它放在哪里。为了防止出现这种错误，需要通过下面的分区方案重新进行分区：

```
create table employees
(
    empno varchar(20) not null,
    empname varchar(20),
    deptno int,
    birthdate date,
    salary int
)
partition by range(salary)(
    partition p0 values less than (5000),
```

```
    partition p1 values less than (10000),
    partition p2 values less than (15000),
    partition p3 values less than (20000),
    partition p4 values less than maxvalue
);
```

maxvalue 表示始终大于最大可能的整数值。salary 列值大于或等于 20000（定义的最高值）的任何行都存储在分区中的 p4 分区。随着数据增大，当员工工资增长到 25000、30000 或更多时，可以使用 alter table 语句为 20000~25000 的工资添加新分区，我们将在第 10 章学习分区的管理知识。

（2）以几乎相同的方式，根据员工部门编号（deptno）对表进行分区。可以使用以下语句来创建数据表的分区：

```
create table employees
(
    empno varchar(20) not null,
    empname varchar(20),
    deptno int,
    birthdate date,
    salary int
)
partition by range(deptno)(
    partition p0 values less than (10),
    partition p1 values less than (20),
    partition p2 values less than (30),
    partition p3 values less than (40)
);
```

在这种分区方案下，所有员工部门编号在 0~10 的数据记录都将存储在 p0 分区中，部门编号在 10~20 的数据记录都将存储在 p1 分区中，以此类推。

（3）也可以根据员工的出生日期（birthdate）进行分区，把出生在同一年的员工信息存储在同一个分区中。

```
create table emp_employees
(
    empno varchar(20) not null,
    empname varchar(20),
    deptno int,
    birthdate date,
    salary int
)
partition by range(year(birthdate)(
    partition p2018 values less than (2018),
    partition p2019 values less than (2019),
    partition p2020 values less than (2020),
    partition p2021 values less than (2021),
    partition pmax values less than maxvalue
);
```

在本分区示例中，我们以 year(birthdate)表达式作为分区依据。

下面以工资范围分区为例，通过测试数据来查询每个分区中分配的数据数量，示例 SQL 语句及其执行结果如下：

```
# 根据工资范围分区
mysql> create table employees
    -> (
    ->        empno varchar(20) not null,
    ->        empname varchar(20),
    ->        deptno int,
    ->        birthdate date,
    ->        salary int
    -> )
    -> partition by range(salary)(
    -> partition p0 values less than (5000),
    -> partition p1 values less than (10000),
    -> partition p2 values less than (15000),
    -> partition p3 values less than (20000),
    -> partition p4 values less than maxvalue
    -> );
Query OK, 0 rows affected (0.10 sec)
# 添加测试数据
mysql> insert into employees (empno,empname,deptno,birthdate,salary) values (1,'wagnwu',1,'2020-06-20',12000);
Query OK, 1 row affected (0.01 sec)
mysql> insert into employees (empno,empname,deptno,birthdate,salary) values (2,'zhangsan',1,'2020-06-20',8000);
Query OK, 1 row affected (0.00 sec)
mysql> insert into employees (empno,empname,deptno,birthdate,salary) values (3,'lisi',1,'2020-06-20',19000);
Query OK, 1 row affected (0.00 sec)
# 查看每个分区中的数据量
mysql> select partition_name as "",table_rows as "" from information_schema.partitions where table_name="employees";
+------+------+
|      |      |
+------+------+
| p0   |    0 |
| p1   |    1 |
| p2   |    1 |
| p3   |    1 |
| p4   |    0 |
+------+------+
5 rows in set (0.00 sec)
```

从结果中可知，p1、p2 和 p3 分区分别有 1 个数据量，p0 和 p4 分区没有数据量。

9.4　列表分区

MySQL 中的列表分区在很多方面类似于范围分区，每个分区都必须明确定义。列表分区与范围分区的区别是：在列表分区中，每个分区是根据列表中的一组值中的列值来定义和选择的，而不是根据一组连续范围的列值来定义和选择的。

例 9.2 使用下面的表结构进行列表分区。

```
create table employees_list
(
    empno varchar(20) not null,
    empname varchar(20),
    deptno int,
    birthdate date,
    salary int
) ;
```

如以上的数据表结构，若想通过部门编号（deptno）来进行列表分区，则可执行如下的 SQL 语句：

```
create table employees_list
(
    empno  varchar(20) not null,
    empname varchar(20),
    deptno  int,
    birthdate date not null,
    salary int
)
partition by list(deptno) (
    partition p0 values in  (10,20,30),
    partition p1 values in  (40,50,60),
    partition p2 values  in  (70,80,90)
);
```

在此分区方案中，部门编号等于 10、20 和 30 的数据存储在 p0 分区中，部门编号等于 40、50 和 60 的数据存储在 p1 分区中，部门编号等于 70、80 和 90 的数据存储在 p2 分区中。

下面用数据进行测试，查询每个分区中分配的数据量，示例 SQL 语句及其执行结果如下：

```
# 创建分区
mysql> create table employees_list
    -> (
    -> empno  varchar(20) not null ,
    -> empname varchar(20),
    -> deptno  int,
    -> birthdate date not null,
    -> salary int
    -> )
    -> partition by list(deptno) (
    -> partition p0 values in  (10,20,30),
    -> partition p1 values in  (40,50,60),
    -> partition p2 values  in  (70,80,90));
Query OK, 0 rows affected (0.06 sec)
# 插入测试数据
mysql> insert into employees_list  (empno,empname,deptno,birthdate,salary) values (1,'zhangsan1',10,'2017-06-20',1998);
Query OK, 1 row affected (0.00 sec)
mysql> insert into employees_list  (empno,empname,deptno,birthdate,salary) values (2,'zhangsan2',50,'2018-06-20',1998);
```

```
Query OK, 1 row affected (0.01 sec)
mysql> insert into employees_list (empno,empname,deptno,birthdate,salary) values
(3,'zhangsan31',20,'2019-06-20',12998);
Query OK, 1 row affected (0.00 sec)
mysql> insert into employees_list (empno,empname,deptno,birthdate,salary) values
(4,'zhangsan32',30,'2021-06-20',12998);
Query OK, 1 row affected (0.00 sec)
# 查看每个分区存储的数据量
mysql> select partition_name as "",table_rows as "" from
information_schema.partitions where table_name="employees_list";
+------+------+
|      |      |
+------+------+
| p0   |    3 |
| p1   |    1 |
| p2   |    0 |
+------+------+
3 rows in set (0.01 sec)
```

由结果可知，p0 分区中有 3 个数据，p1 分区中有 1 个数据，p2 分区中有 0 个数据。

在列表分区的方案中，如果插入的数据中分区字段的值不在分区列表中，则会出现错误提示信息，示例如下：

```
mysql> insert into employees_list(empno,empname,deptno,birthdate,salary) values
(5,'zhangsan5',100,'2021-06-20',12998);
ERROR 1526 (HY000): Table has no partition for value 100
```

如上所示，插入一条员工信息，该员工部门编号是 100，此编号并不在分区的数值列表中，因此提示不知道应将该条信息存储在哪个分区中。所以，如果要在一条语句中添加多条数据，则可以使用 ignore 关键字忽略这个错误。

例 9.3 添加数据时，使用 ignore 关键字忽略部分数据错误，示例 SQL 语句及其执行结果如下：

```
# 插入操作，忽略部分数据错误
mysql> insert ignore into employees_list (empno,empname,deptno,birthdate,salary)
values(6,'zhangsan6',10,'2021-06-20',
12998),(7,'zhangsan7',100,'2021-06-20',12998);
Query OK, 1 row affected, 1 warning (0.00 sec)
Records: 2  Duplicates: 1  Warnings: 1
# 查询结果显示，不在分区列表中的数据没有添加进去
mysql> select * from employees_list;
+-------+------------+--------+------------+--------+
| empno | empname    | deptno | birthdate  | salary |
+-------+------------+--------+------------+--------+
|     1 | zhangsan1  |     10 | 2017-06-20 |   1998 |
|     3 | zhangsan31 |     20 | 2019-06-20 |  12998 |
|     4 | zhangsan32 |     30 | 2021-06-20 |  12998 |
|     6 | zhangsan6  |     10 | 2021-06-20 |  12998 |
|     2 | zhangsan2  |     50 | 2018-06-20 |   1998 |
+-------+------------+--------+------------+--------+
5 rows in set (0.00 sec)
```

9.5 列 分 区

列（column）分区是范围分区和列表分区的变体。列分区允许在分区键中使用多个列。它分为两种分区类型，一种是范围列（range column）分区，另一种是列表列（list column）分区。列分区支持使用非整数列来定义值范围或列表成员。支持的数据类型说明如下：

- 支持 tinyint、smallint、mediumint、int 和 bigint 类型的列作为分区列。
- 不支持 decimal 或 float 类型的列作为分区列。
- 支持 date 和 datetime 类型的列作为分区列。
- 支持 char、varchar、binary 和 varbinary 类型的列作为分区列。
- 不支持 text 和 blob 类型的列作为分区列。

9.5.1 范围列分区

范围列分区类似于范围分区，不一样的是范围列分区允许使用多个列值的范围来定义分区。此外，还可以使用整数类型以外的类型列来定义范围。范围列分区和范围分区在以下方面也有很大不同：

- 范围列分区不接收表达式，只接收列名。
- 范围列分区接收一列或多列的列表。
- 范围列分区基于元组（列值列表）之间的比较，而不是标量值之间的比较。
- 范围列分区的列不限于整数列，字符串、date 以及 datetime 类型的列也可以作为分区列。

范围列分区创建分区表的基本语法如下：

```
create table table_name
partitioned by range columns(column_list) (
    partition partition_name values less than (value_list)[,
    partition partition_name values less than (value_list)][,
    ...]
)
column_list:
    column_name[, column_name][, ...]
value_list:
    value[, value][, ...]
```

在上述语法中，column_list 是一列或多列的列表（分区列的列表）；value_list 是值列表（即分区定义的值列表），而且 value_list 必须为每个分区提供定义。每个 value_list 的值数量与 column_list 的列数量相同，也就是说，如果用户在 columns 子句中使用 N 个列，那么每个 value less than 子句就必须提供一个具有 N 个值的列表。

分区列的列表中的元素和定义每个分区的值列表中的元素必须以相同的顺序出现。此外，值列表中的每个元素都必须与分区列的列表中的相应元素的数据类型相同。但是，分区列的列表和值列表中列名的顺序不必与 create table 语句中定义表列的顺序相同。下面是一个简单的范围列分区示例。

例 9.4 范围列分区示例的 SQL 语句及其执行结果如下：

```
mysql> create table rtable(
    ->    a int,
    ->    b int,
    ->    c char(8),
    ->    d int
    -> )
    -> partition by range columns(a,d,c) (
    ->    partition p0 values less than (5,20,'aa'),
    ->    partition p1 values less than (10,30,'cc'),
    ->    partition p2 values less than (15,80,'dd'),
    ->    partition p3 values less than (maxvalue,maxvalue,maxvalue)
    -> );
Query OK, 0 rows affected (0.12 sec)
```

在上述示例中，表 rtable 包含 4 列（a、b、c、d 列），该 columns 子句的分区列表使用了其中 3 列，其顺序为 a、d、c，并且每个值列表元组都具有(int,int,char(8))形式，对应列 a、d 和 c（按该顺序）使用的数据类型。

把与 columns 子句中分区列的列表相匹配的元组与 value less than 子句中用于定义表分区的元组进行比较，就可以确定将插入的行放置到具体哪个分区中。因为我们比较的是元组而不是标量值，所以 value less than 用于范围列分区的语义与使用简单范围分区的语义有所不同。接下来用两个示例来展示这两种不同分区的效果。

1）范围分区与使用多个分区列进行范围列分区的区别

例 9.5 范围分区与使用多个分区列进行范围列分区的区别示例。

（1）使用范围分区创建表 tablea，SQL 语句如下：

```
create table tablea (
   a int,
   b int
)
partition by range (a) (
   partition p0 values less than (5),
   partition p1 values less than (maxvalue)
);
# 在表中插入3行，每一行的列值 a 都是5，那么3行全部存储在 p1 分区中
mysql> insert into tablea values (5,10), (5,11), (5,12);
Query OK, 3 rows affected (0.00 sec)
Records: 3  Duplicates: 0  Warnings: 0
# 查询各个分区的数据分布
mysql> select partition_name,table_rows
    ->    from information_schema.partitions
    ->    where table_name = 'tablea';
+----------------+------------+
| partition_name | table_rows |
```

```
+----------------+------------+
| p0             |          0 |
| p1             |          3 |
+----------------+------------+
2 rows in set (0.00 sec)
```

由结果可知，3 行数据全部存储在 p1 分区中。

（2）创建一个类似于 tablea 的表，使用范围列对列 a 和列 b 创建分区，SQL 语句如下：

```
create table tablea(
    a int,
    b int
)
partition by range columns(a, b) (
    partition p0 values less than (5, 12),
    partition p1 values less than (maxvalue, maxvalue)
);
# 插入与 tablea 刚刚插入的行完全相同的行，tablea 的行分布会大不相同
mysql> insert into tableavalues (5,10), (5,11), (5,12);
Query OK, 3 rows affected (0.00 sec)
Records: 3  Duplicates: 0  Warnings: 0
# 查询每个分区的数据分布
mysql> select partition_name,table_rows
    -> from information_schema.partitions
    -> where table_name = 'rc1';
+--------------+----------------+------------+
| table_schema | partition_name | table_rows |
+--------------+----------------+------------+
| p            | p0             |          2 |
| p            | p1             |          1 |
+--------------+----------------+------------+
2 rows in set (0.00 sec)
```

由结果可知，有 2 行数据存储在 p0 分区中，有 1 行数据存储在 p1 分区中。

出现上面问题的原因是我们比较的是行值而不是标量值。我们把将插入的行值与在 tablea 中定义 p0 分区的限制行值进行比较，结果如下：

```
mysql> select (5,10) < (5,12), (5,11) < (5,12), (5,12) < (5,12);
+-----------------+-----------------+-----------------+
| (5,10) < (5,12) | (5,11) < (5,12) | (5,12) < (5,12) |
+-----------------+-----------------+-----------------+
|               1 |               1 |               0 |
+-----------------+-----------------+-----------------+
1 row in set (0.00 sec)
```

因此(5,10)和(5,11)会存储在 p0 分区，(5,12)会存储在 p1 分区。

2）范围分区与仅使用单个分区列进行范围列分区的区别

对于仅使用单个分区列进行范围列分区的数据表，分区中行的存储与按范围分区的行的存储效果是相同的。

例 9.6 范围分区与使用单个分区列进行范围列分区的区别，示例 SQL 语句及其执行结

果如下：

```
# 创建一个使用单个分区列进行范围列分区的数据表
create table tableb(
    a int,
    b int
)
partition by range columns (a)  (
    partition p0 values less than (5),
    partition p1 values less than (maxvalue)
);
# 把行(5,10)、(5,11)和(5,12)插入这张表中
mysql> insert into tableb values (5,10), (5,11),(5,12);
Query OK, 3 rows affected (0.00 sec)
Records: 3  Duplicates: 0  Warnings: 0
# 查询分区的数据分布，可以看到它们的位置与之前使用范围分区创建和填充的表相同
mysql> select partition_name,table_rows
    ->     from information_schema.partitions
    ->     where table_name = 'tableb';
+---------------+----------------+------------+
| table_schema  | partition_name | table_rows |
+---------------+----------------+------------+
| p             | p0             |          0 |
| p             | p1             |          3 |
+---------------+----------------+------------+
2 rows in set (0.00 sec)
```

需要注意的是，范围列的一个或多个列的限制值在连续分区定义中可以重复，只要保证用于定义分区的列值元组是严格递增的即可。例如，以下分区语句都是正确的：

```
create table tablec(
    a int,
    b int
)
partition by range columns(a,b) (
    partition p0 values less than (0,10),
    partition p1 values less than (10,20),
    partition p2 values less than (10,30),
    partition p3 values less than (maxvalue,maxvalue)
);
create table tablec(
    a int,
    b int
)
partition by range columns(a,b) (
    partition p0 values less than (0,10),
    partition p1 values less than (10,20),
    partition p2 values less than (10,30),
    partition p3 values less than (10,35),
    partition p4 values less than (20,40),
    partition p5 values less than (maxvalue,maxvalue)
);
create table tabled(
    a int,
    b int,
```

```
    c int
)
partition by range columns(a,b,c) (
    partition p0 values less than (0,25,50),
    partition p1 values less than (10,20,100),
    partition p2 values less than (10,30,50),
    partition p3 values less than (maxvalue,maxvalue,maxvalue)
);
```

3）范围列可以使用 MySQL 客户端所需的元组来测试连续的分区定义

例 9.7 在设计分区时，范围列使用 MySQL 客户端所需的元组来测试连续的分区定义，SQL 语句如下：

```
mysql> select (0,25,50) < (10,20,100), (10,20,100) < (10,30,50);
+-------------------------+--------------------------+
| (0,25,50) < (10,20,100) | (10,20,100) < (10,30,50) |
+-------------------------+--------------------------+
|                       1 |                        1 |
+-------------------------+--------------------------+
1 row in set (0.00 sec)
```

由结果可知，分区定义是连续的。

如果 create table 语句包含非严格递增顺序的分区定义，那么语句会执行失败并显示错误提示信息，示例 SQL 语句及其执行结果如下：

```
mysql> create table rcf (
    ->     a int,
    ->     b int,
    ->     c int
    -> )
    -> partition by range columns(a,b,c) (
    ->     partition p0 values less than (0,25,50),
    ->     partition p1 values less than (20,20,100),
    ->     partition p2 values less than (10,30,50),
    ->     partition p3 values less than (maxvalue,maxvalue,maxvalue)
    -> );
ERROR 1493 (HY000): value less than value must be strictly increasing for each partition
```

当遇到此类错误时，可以使用 MySQL 客户端所需的元组来测试连续的分区定义，示例 SQL 语句及其执行结果如下：

```
mysql> select (0,25,50) < (20,20,100), (20,20,100) < (10,30,50);
+-------------------------+--------------------------+
| (0,25,50) < (20,20,100) | (20,20,100) < (10,30,50) |
+-------------------------+--------------------------+
|                       1 |                        0 |
+-------------------------+--------------------------+
1 row in set (0.00 sec)
```

由结果可知，分区定义不连续。

4)范围列分区可以使用非整数列作为分区列

如前所述,范围列也可以使用非整数列作为分区列。

例 9.8　创建一张名为 employees 的数据表,对表中的 lname 列进行范围列分区,示例 SQL 语句如下:

```
# 创建名为 employees 的数据表
create table employees (
    id int not null,
    fname varchar(30),
    lname varchar(30),
    hired date not null default '1970-01-01',
    separated date not null default '9999-12-31',
    job_code int not null,
    store_id int not null
);
# 使用范围列分区创建此表的一个版本,将每一行数据存储在基于员工姓氏的 4 个分区之一
partition by range columns (lname)  (
    partition p0 values less than ('g'),
    partition p1 values less than ('m'),
    partition p2 values less than ('t'),
    partition p3 values less than (maxvalue)
);
# 也可以通过执行以下 alter table 语句对先前创建的 employees 表进行分区
alter table employees partition by range columns (lname)  (
    partition p0 values less than ('g'),
    partition p1 values less than ('m'),
    partition p2 values less than ('t'),
    partition p3 values less than (maxvalue)
);
```

9.5.2　列表列分区

MySQL 8 提供对列表列分区的支持。列表列分区是列表分区的一种变体,允许使用多个列作为分区键。

1)使用 city 列通过列表列分区创建一张数据表

例 9.9　一家企业在 12 个省市拥有客户,出于销售和营销的目的,需要将这 12 个省市组织成 4 个区域,每个区域 3 个省市,如表 9-1 所示。

表 9-1　客户数据

地区	省市
1	河南省、湖北省、湖南省
2	广东省、广西壮族自治区、海南省
3	上海市、江苏省、浙江省
4	北京市、天津市、河北省

通过列表列分区为客户数据创建一张表,该表根据客户所在省市的名称将一条记录(对

应一行的数据）分配给这些区域对应的 4 个分区中的一个，示例 SQL 语句如下：

```
create table customers (
    name varchar(25),
    street_1 varchar(30),
    street_2 varchar(30),
    city varchar(15),
    renewal date
)
partition by list columns(city) (
    partition pregion_1 values in('河南省', '湖北省', '湖南省'),
    partition pregion_2 values in('广东省', '广西壮族自治区', '海南省'),
    partition pregion_3 values in('上海市', '江苏省', '浙江省'),
    partition pregion_4 values in('北京市', '天津市', '河北省')
);
```

2）使用 date 和 datetime 数据类型的列通过列表列分区创建一张表

例 9.10 根据表中的 renewal 列进行列表列分区，并将表分为 4 个分区，示例 SQL 语句如下：

```
create table customers (
    first_name varchar(25),
    last_name varchar(25),
    street_1 varchar(30),
    street_2 varchar(30),
    city varchar(15),
    renewal date
)
partition by list columns(renewal) (
    partition pweek_1 values in('2021-02-01', '2021-02-02', '2021-02-03',
        '2021-02-04', '2021-02-05', '2021-02-06', '2021-02-07'),
    partition pweek_2 values in('2021-02-08', '2021-02-09', '2021-02-10',
        '2021-02-11', '2021-02-12', '2021-02-13', '2021-02-14'),
    partition pweek_3 values in('2021-02-15', '2021-02-16', '2021-02-17',
        '2021-02-18', '2021-02-19', '2021-02-20', '2021-02-21'),
    partition pweek_4 values in('2021-02-22', '2021-02-23', '2021-02-24',
        '2021-02-25', '2021-02-26', '2021-02-27', '2021-02-28')
);
```

如果涉及的日期数量非常大，定义与维护就会变得很麻烦，在这种情况下，建议使用范围分区或范围列分区。

例 9.11 将 date 数据类型的列作为分区键进行范围列分区，示例 SQL 语句如下：

```
create table customers(
    first_name varchar(25),
    last_name varchar(25),
    street_1 varchar(30),
    street_2 varchar(30),
    city varchar(15),
    renewal date
)
partition by range columns(renewal) (
    partition pweek_1 values less than('2021-02-09'),
```

```
    partition pweek_2 values less than('2021-02-15'),
    partition pweek_3 values less than('2021-02-22'),
    partition pweek_4 values less than('2021-03-01')
);
```

9.6 哈希分区

1. 常规哈希分区

哈希分区主要用来确保数据在预先确定的分区中平均分布。在范围分区和列表分区中必须明确指定每个分区对应的数据；而在哈希分区中，MySQL 会自动完成这些工作，用户所要做的只是为将要被哈希的列值指定一个表达式以及表分区的数量。

要对表进行哈希分区，必须在 create table 语句后附加一个子句，这个子句可以是一个返回整数的表达式，也可以是 MySQL 整数类型列的名称。

例 9.12 创建一张数据表，根据表中的 store_id 列进行哈希分区并分为 4 个分区，示例 SQL 语句如下：

```
create table employees (
    id int not null,
    fname varchar(30),
    lname varchar(30),
    hired date not null default '1970-01-01',
    separated date not null default '9999-12-31',
    job_code int,
    store_id int
)
partition by hash(store_id)
partitions 4;
```

如果分区语句不包含 partition 子句，则分区数默认为 1；如果分区语句包含 partition 子句，则必须在后面指定分区的数量，否则会提示语法错误。

还可以使用为 SQL 返回整数的表达式：

```
create table employees (
    id int not null,
    fname varchar(30),
    lname varchar(30),
    hired date not null default '1970-01-01',
    separated date not null default '9999-12-31',
    job_code int,
    store_id int
)
partition by hash( year(hired) )
partitions 4;
```

完整示例的 SQL 语句及其执行结果如下：

```
# 创建分区表
mysql> create table emp
    -> (
    -> empno varchar(20) not null,
    -> empname varchar(20),
    -> deptno int,
    -> birthdate date not null,
    -> salary int
    -> ) partition by hash(year(birthdate))
    -> partitions 4;
Query OK, 0 rows affected (0.07 sec)
# 插入测试数据
mysql> insert into emp (empno,empname,deptno,birthdate,salary) values (1,'zhangsan1',10,'2017-06-20',998);
insert into emp (empno,empname,deptno,birthdate,salary) values (2,'zhangsan2',10,'2018-06-20',1998);
insert into emp (empno,empname,deptno,birthdate,salary) values (3,'zhangsan31',20,'2019-06-20',2998);
insert into emp (empno,empname,deptno,birthdate,salary) values (4,'zhangsan32',30,'2020-06-20',2998);Query OK, 1 row affected (0.00 sec)
mysql> insert into emp (empno,empname,deptno,birthdate,salary) values (2,'zhangsan2',10,'2018-06-20',1998);
Query OK, 1 row affected (0.01 sec)
mysql> insert into emp (empno,empname,deptno,birthdate,salary) values (3,'zhangsan31',20,'2019-06-20',2998);
Query OK, 1 row affected (0.01 sec)
mysql> insert into emp (empno,empname,deptno,birthdate,salary) values (4,'zhangsan32',30,'2020-06-20',2998);
Query OK, 1 row affected (0.00 sec)
# 查询分区 p0 中的数据
mysql> select * from emp partition(p0);
+-------+------------+--------+------------+--------+
| empno | empname    | deptno | birthdate  | salary |
+-------+------------+--------+------------+--------+
| 4     | zhangsan32 |     30 | 2020-06-20 |   2998 |
+-------+------------+--------+------------+--------+
1 row in set (0.00 sec)
# 查询分区 p1 中的数据
mysql> select * from emp partition(p1);
+-------+-----------+--------+------------+--------+
| empno | empname   | deptno | birthdate  | salary |
+-------+-----------+--------+------------+--------+
| 1     | zhangsan1 |     10 | 2017-06-20 |    998 |
+-------+-----------+--------+------------+--------+
1 row in set (0.01 sec)
# 查询分区 p2 中的数据
mysql> select * from emp partition(p2);
+-------+-----------+--------+------------+--------+
| empno | empname   | deptno | birthdate  | salary |
+-------+-----------+--------+------------+--------+
| 2     | zhangsan2 |     10 | 2018-06-20 |   1998 |
+-------+-----------+--------+------------+--------+
1 row in set (0.00 sec)
# 查询分区 p3 中的数据
mysql> select * from emp partition(p3);
+-------+------------+--------+------------+--------+
```

```
| empno | empname   | deptno | birthdate  | salary |
+-------+-----------+--------+------------+--------+
| 3     | zhangsan31|   20   | 2019-06-20 | 2998   |
+-------+-----------+--------+------------+--------+
1 row in set (0.00 sec)
```

2. 线性哈希分区

MySQL 还支持线性哈希，它与常规哈希的不同之处在于：线性哈希使用线性二次幂算法，而常规哈希使用哈希函数值的模数。在语法上，线性哈希分区唯一区别于常规哈希的地方是在 partition by 子句中添加了 linear 关键字。

例 9.13 线性哈希分区的示例 SQL 语句如下：

```
create table employees (
    id int not null,
    fname varchar(30),
    lname varchar(30),
    hired date not null default '1970-01-01',
    separated date not null default '9999-12-31',
    job_code int,
    store_id int
)
partition by linear hash( year(hired) )
partitions 4;
```

线性哈希分区数 N 是根据以下算法导出的：

假设要保存记录的分区编号为 n，num 为一个非负整数，表示分割成的分区的数量，那么 N 可以通过以下步骤得到：

步骤 01 找到一个大于或等于 num 的 2 的幂，这个值为 v，v 可以通过公式：v = power(2, ceiling(log(2, num))) 得到。

步骤 02 设置 n=f(column_list) & (v–1)。

步骤 03 当 n≥num 时，设置 v=ceiling(V/2)，n=n&(v–1)，只要满足 n≥num，就一直循环执行。

例 9.14 假设 t1 使用线性哈希分区，并且具有 6 个分区的表 table 是使用以下语句创建的：

```
create table t1 (col1 int, col2 char(5), col3 date)
    partition by linear hash( year(col3) )
partitions 6;
```

现在要插入两条记录，使 t1 具有 col3 列值'2003-04-14'和'1998-10-19'。其中第一条记录的分区数计算如下：

```
V = power(2, ceiling( log(2,6) )) = 8
N = year('2003-04-14') & (8 - 1)
  = 2003 & 7
  = 3
(3 ≥ 6 不成立：数据在第 3 个分区中)
```

存储第二条记录的分区数计算如下：

```
V = 8
N = year('1998-10-19') & (8 - 1)
  = 1998 & 7
  = 6
(6 ≥ 6 成立：还需要再次计算)
N = 6 & ((8 / 2) - 1)
  = 6 & 3
  = 2
(2 ≥ 6 不成立：数据存储在第 2 个分区中)
```

9.7 键 分 区

键分区将表中的数据按照特定的键值进行分区。在键分区中，每个分区都包含相同键值的数据，不同键值的数据则存储在不同的分区中。

键分区和哈希分区很相似，但是键分区支持除 text 和 blob 类型之外的所有数据类型的列，而哈希分区只支持数字类型的列；键分区不允许使用用户自定义的表达式进行分区，而是使用系统提供的哈希函数进行分区。当表中存在主键或者唯一键时，如果创建键分区时没有指定列，则系统默认会选择主键列作为分区列；如果不存在主键列，则会选择非空的唯一键列作为分区列。

> **提 示**
> 唯一列作为分区列时，唯一列不能为 null。

例 9.15 键分区示例的 SQL 语句如下：

```
mysql> create table tb_key (
    ->     id int,
    ->     var char(32)
    -> )
    -> partition by key(var)
    -> partitions 10;
Query OK, 0 rows affected (0.05 sec)
# 查询当前表的分区情况
mysql> select partition_name,partition_method,partition_expression,
partition_description,table_rows,subpartition_name,subpartition_method,subpartition
_expression
    -> from information_schema.partitions where table_schema=schema()
    -> and table_name='tb_key';
+----------------+------------------+----------------------+--------------------
----+------------+-------------------+---------------------+----------------------
--+
| PARTITION_NAME | PARTITION_METHOD | PARTITION_EXPRESSION | PARTITION_DESCRIPTION
| TABLE_ROWS | SUBPARTITION_NAME | SUBPARTITION_METHOD | SUBPARTITION_EXPRESSION |
+----------------+------------------+----------------------+--------------------
----+------------+-------------------+---------------------+----------------------
--+
| p9             | KEY              | `var`                | NULL                 |
```

```
 0 | NULL           | NULL             | NULL                |                    |
 | p8             | KEY              | `var`               | NULL               |
 0 | NULL           | NULL             | NULL                |                    |
 | p7             | KEY              | `var`               | NULL               |
 0 | NULL           | NULL             | NULL                |                    |
 | p6             | KEY              | `var`               | NULL               |
 0 | NULL           | NULL             | NULL                |                    |
 | p5             | KEY              | `var`               | NULL               |
 0 | NULL           | NULL             | NULL                |                    |
 | p4             | KEY              | `var`               | NULL               |
 0 | NULL           | NULL             | NULL                |                    |
 | p3             | KEY              | `var`               | NULL               |
 0 | NULL           | NULL             | NULL                |                    |
 | p2             | KEY              | `var`               | NULL               |
 0 | NULL           | NULL             | NULL                |                    |
 | p1             | KEY              | `var`               | NULL               |
 0 | NULL           | NULL             | NULL                |                    |
 | p0             | KEY              | `var`               | NULL               |
 0 | NULL           | NULL             | NULL                |                    |
 +----------------+------------------+---------------------+--------------------+
10 rows in set (0.00 sec)
```

本例根据 var 列通过键分区把 tb_key 表分为了 10 个分区。

9.8 子 分 区

子分区也称为复合分区，是对分区表中每个分区的进一步划分。子分区可以分为范围-哈希复合分区、范围-键复合分区、列表-哈希复合分区、列表-键复合分区。

1．范围-哈希复合分区

范围-哈希（range-hash）复合分区示例如下：

例 9.16　对 salary 列进行范围分区，并对 birthdate 列进行哈希分区，示例 SQL 语句如下：

```
create table emp(
    empno varchar(20) not null,
    empname varchar(20),
    deptno int,
    birthdate date not null,
    salary int
)
partition by range(salary)
subpartition by hash(year(birthdate))
subpartitions 3(
    partition p1 values less than (2000),
    partition p2 values less than maxvalue
);
```

本例首先创建了一张名为 emp 的表，表中包含 5 列：empno、empname、deptno、birthdate 和 salary。然后按照 salary 列的薪资范围将表进行分区，并对 birthdate 列采用 year 进行哈希分区，子分区数为 3。在此分区方案中，将数据分成两个范围分区 p1 和 p2，每个范围分区又分为 3 个子分区，其中 p1 分区存储 salary 小于 2000 的值，p2 分区存储所有大于或等于 maxvalue 的值。

2. 范围-键复合分区

范围-键（range-key）复合分区示例如下：

例 9.17 对 salary 列进行范围分区，并对 birthdate 列进行键分区，示例 SQL 语句如下：

```
create table emp
(
    empno varchar(20) not null,
    empname varchar(20),
    deptno int,
    birthdate date not null,
    salary int
)
partition by range(salary)
subpartition by key(birthdate)
subpartitions 3
(
    partition p1 values less than (2000),
    partition p2 values less than maxvalue
);
```

本例对 salary 列进行范围分区，并对 birthdate 列进行键分区，子分区数为 3。本例语句与例 9.16 的不同之处在于对出生日期进行了 key 分区，这意味着根据出生日期，相同的值将进入同一个分区，这有助于减少查询时需要扫描的分区数量。

3. 列表-哈希复合分区

列表-哈希（list-hash）复合分区示例如下：

例 9.18 对 deptno 列进行列表分区，并对 birthdate 列进行哈希分区，示例 SQL 语句如下：

```
create table emp (
    empno varchar(20) not null,
    empname varchar(20),
    deptno int,
    birthdate date not null,
    salary int
)
partition by list (deptno)
subpartition by hash(year(birthdate))
subpartitions 3
(
    partition p1 values in  (10),
```

```
    partition p2 values in  (20)
);
```

本例对 deptno 列进行列表分区，并采用 year(birthdate)的哈希算法进行子分区，子分区数为 3。在子分区中，使用出生年份数字的哈希值来对数据进行子分区，这个哈希值将把同一年份出生的员工放在同一个子分区中。在此分区方案中，将数据分成两个列表分区 p1 和 p2，每个列表分区又分为 3 个子分区，其中 p1 分区中存储部门编号为 10 的员工，p2 分区中存储部门编号为 20 的员工。

4. 列表-键复合分区

列表-键（list-key）复合分区示例如下：

例 9.19　对 deptno 列进行列表分区，并对 birthdate 列进行键分区，示例 SQL 语句如下：

```
create table emp (
    empno varchar(20) not null,
    empname varchar(20),
    deptno int,
    birthdate date not null,
    salary int
)
partition by list (deptno)
subpartition by key(birthdate)
subpartitions 3
(
    partition p1 values in  (10),
    partition p2 values in  (20)
);
```

与例 9.18 的 SQL 语句不同的是，本例使用 subpartition by key(birthdate)来定义子分区键，而不是哈希算法。因此，子分区的策略是按照员工的出生日期进行分区。在此分区方案中，将数据分成了两个列表分区 p1 和 p2，然后每个列表分区又划分为 3 个子分区。

9.9　MySQL 分区对 null 的处理

MySQL 中的分区不会禁止 null 作为分区表达式的值，无论列值还是用户提供的表达式的值，都允许 null 用作必须产生整数的表达式的值。需要注意的是，null 不是数字。MySQL 中的分区将 null 视为小于任何非 null 值，这意味着 null 在不同类型的分区中的处理方式也不同。本节将学习每种 MySQL 分区类型中存储 null 的数据行在确定分区时是如何处理 null 的，并为每种分区提供示例。

1. 使用范围分区处理 null

要将一行数据插入分区中，如果用于确定范围分区的列值为 null，那么该行将插入最低分区中。

例 9.20　使用范围分区处理 null，示例 SQL 语句及其执行结果如下：

```
mysql> create table t1 (
    ->     c1 int,
    ->     c2 varchar(20)
    -> )
    -> partition by range(c1) (
    ->     partition p0 values less than (0),
    ->     partition p1 values less than (10),
    ->     partition p2 values less than maxvalue
    -> );
Query OK, 0 rows affected (0.09 sec)

mysql> create table t2 (
    ->     c1 int,
    ->     c2 varchar(20)
    -> )
    -> partition by range(c1) (
    ->     partition p0 values less than (-5),
    ->     partition p1 values less than (0),
    ->     partition p2 values less than (10),
    ->     partition p3 values less than maxvalue
    -> );
Query OK, 0 rows affected (0.09 sec)
```

查询这两条语句创建的分区 information_schema：

```
mysql> select table_name, partition_name, table_rows, avg_row_length, data_length
    >   from information_schema.partitions
    >   where table_schema = 'p' and table_name like 't_';
+------------+----------------+------------+----------------+-------------+
| table_name | partition_name | table_rows | avg_row_length | data_length |
+------------+----------------+------------+----------------+-------------+
| t1         | p0             |          0 |              0 |           0 |
| t1         | p1             |          0 |              0 |           0 |
| t1         | p2             |          0 |              0 |           0 |
| t2         | p0             |          0 |              0 |           0 |
| t2         | p1             |          0 |              0 |           0 |
| t2         | p2             |          0 |              0 |           0 |
| t2         | p3             |          0 |              0 |           0 |
+------------+----------------+------------+----------------+-------------+
7 rows in set (0.00 sec)
# 插入数据，给分区字段赋值为 null
mysql> insert into t1 values (null, 'mothra');
Query OK, 1 row affected (0.00 sec)

mysql> insert into t2 values (null, 'mothra');
Query OK, 1 row affected (0.00 sec)

mysql> select * from t1;
```

```
+------+--------+
| id   | name   |
+------+--------+
| null | mothra |
+------+--------+
1 row in set (0.00 sec)

mysql> select * from t2;
+------+--------+
| id   | name   |
+------+--------+
| null | mothra |
+------+--------+
1 row in set (0.00 sec)
```

查询两个表中分区的数据量：

```
mysql> select table_name, partition_name, table_rows, avg_row_length, data_length
    >  from information_schema.partitions
    >  where table_schema = 'p' and table_name like 't_';
+------------+----------------+------------+----------------+-------------+
| table_name | partition_name | table_rows | avg_row_length | data_length |
+------------+----------------+------------+----------------+-------------+
| t1         | p0             |          1 |             20 |          20 |
| t1         | p1             |          0 |              0 |           0 |
| t1         | p2             |          0 |              0 |           0 |
| t2         | p0             |          1 |             20 |          20 |
| t2         | p1             |          0 |              0 |           0 |
| t2         | p2             |          0 |              0 |           0 |
| t2         | p3             |          0 |              0 |           0 |
+------------+----------------+------------+----------------+-------------+
7 rows in set (0.01 sec)
```

通过结果可以看出，null 数据被存储在编号最小的分区中。

下面删除这些分区，然后重新运行 select 语句来证明 null 数据行存储在每张表的编号最小的分区中：

```
mysql> alter table t1 drop partition p0;
Query OK, 0 rows affected (0.16 sec)

mysql> alter table t2 drop partition p0;
Query OK, 0 rows affected (0.16 sec)

mysql> select * from t1;
Empty set (0.00 sec)

mysql> select * from t2;
Empty set (0.00 sec)
```

首先使用 alter table 命令分别删除 t1 和 t2 表中的分区 p0；然后，分别使用 select 命令查询 t1 和 t2 表的内容。由于这些表中没有数据，因此查询结果为空集。由此证明 null 数据行存储在每张表的编号最小的分区中。

SQL 函数的分区表达式也是以这种方式来处理 null 数据的。

例 9.21 若 SQL 函数返回 null，则该行存储在编号最小的分区中，示例 SQL 语句如下：

```
create table t3 (
    id int,
    dt date
)
partition by range( year(dt) ) (
    partition p0 values less than (1990),
    partition p1 values less than (2000),
    partition p2 values less than maxvalue
);
```

与其他 MySQL 函数一样，year(null)返回 null，所以被插入 p0 分区。

2. 使用列表分区处理 null

当且仅当使用包含的值列表定义其分区之一时，列表分区的数据表才允许 null 值插入分区。

例 9.22 将不在列表分区定义范围内的 null 值插入列表分区的数据表中，示例 SQL 语句及其执行结果如下：

```
mysql> create table ts1 (
    ->     c1 int,
    ->     c2 varchar(20)
    -> )
    -> partition by list(c1) (
    ->     partition p0 values in (0, 3, 6),
    ->     partition p1 values in (1, 4, 7),
    ->     partition p2 values in (2, 5, 8)
    -> );
Query OK, 0 rows affected (0.01 sec)

mysql> insert into ts1 values (9, 'mothra');
ERROR 1504 (HY000): Table has no partition for value 9

mysql> insert into ts1 values (null, 'mothra');
ERROR 1504 (HY000): Table has no partition for value null
```

在本例中，只能将 c1 列值介于 0～8 的行数据插入 ts1 表中。由于列值 9 和 null 不在列表分区定义的范围内，因此插入时报错。

例 9.23 将列表分区定义范围内的 null 值插入列表分区的数据表中，示例 SQL 语句及其执行结果如下：

```
# 创建表 ts2 和 ts3，表中的分区包含列值为 null 的行数据
mysql> create table ts2 (
    ->     c1 int,
    ->     c2 varchar(20)
    -> )
    -> partition by list(c1) (
    ->     partition p0 values in (0, 3, 6),
```

```
    ->     partition p1 values in (1, 4, 7),
    ->     partition p2 values in (2, 5, 8),
    ->     partition p3 values in (null)
    -> );
Query OK, 0 rows affected (0.01 sec)

mysql> create table ts3 (
    ->     c1 int,
    ->     c2 varchar(20)
    -> )
    -> partition by list(c1) (
    ->     partition p0 values in (0, 3, 6),
    ->     partition p1 values in (1, 4, 7, null),
    ->     partition p2 values in (2, 5, 8)
    -> );
Query OK, 0 rows affected (0.01 sec)
mysql> insert into ts2 values (null, 'mothra');
Query OK, 1 row affected (0.00 sec)
mysql> insert into ts3 values (null, 'mothra');
Query OK, 1 row affected (0.00 sec)
# 查询每个分区中的数据分布数量
mysql> select table_name, partition_name, table_rows, avg_row_length, data_length
    >  from information_schema.partitions
    >  where table_schema = 'p' and table_name like 'ts_';
+------------+----------------+------------+----------------+-------------+
| table_name | partition_name | table_rows | avg_row_length | data_length |
+------------+----------------+------------+----------------+-------------+
| ts2        | p0             |          0 |              0 |           0 |
| ts2        | p1             |          0 |              0 |           0 |
| ts2        | p2             |          0 |              0 |           0 |
| ts2        | p3             |          1 |             20 |          20 |
| ts3        | p0             |          0 |              0 |           0 |
| ts3        | p1             |          1 |             20 |          20 |
| ts3        | p2             |          0 |              0 |           0 |
+------------+----------------+------------+----------------+-------------+
7 rows in set (0.01 sec)
```

如以上结果所示，值为 null 的行数据被插入列表分区对应的分区中。

3. 使用哈希分区和键分区处理 null

对于哈希分区或者键分区的表而言，处理值为 null 的行数据的方式与列表分区、范围分布略有不同。在哈希分区和键分区的表中，任何产生 null 值的分区表达式的返回值都为 0。

例 9.24 在哈希分区中插入值为 null 的列值，示例 SQL 语句及其执行结果如下：

```
# 创建一张数据表，使用哈希分区
mysql> create table th (
    ->     c1 int,
    ->     c2 varchar(20)
    -> )
    -> partition by hash(c1)
    -> partitions 2;
Query OK, 0 rows affected (0.00 sec)
```

查询属于该表的分区
```
mysql>select
>   table_name,partition_name,table_rows,avg_row_length,data_length
    >   from information_schema.partitions
    >   where table_schema = 'p' and table_name ='th';
+------------+----------------+------------+----------------+-------------+
| table_name | partition_name | table_rows | avg_row_length | data_length |
+------------+----------------+------------+----------------+-------------+
| th         | p0             |          0 |              0 |           0 |
| th         | p1             |          0 |              0 |           0 |
+------------+----------------+------------+----------------+-------------+
2 rows in set (0.00 sec)
```

从查询结果可知,每个分区存储的数据条数都是 0。现在插入两行数据,其 c1 列值为 null,并验证这些行的数据是否已插入:

```
mysql> insert into th values (null, 'mothra'), (0, 'gigan');
Query OK, 1 row affected (0.00 sec)
mysql> select * from th;
+------+---------+
| c1   | c2      |
+------+---------+
| null | mothra  |
+------+---------+
|    0 | gigan   |
+------+---------+
2 rows in set (0.01 sec)
```

查询每个分区中数据的分布情况:

```
mysql> select table_name, partition_name, table_rows, avg_row_length, data_length
    >   from information_schema.partitions
    >   where table_schema = 'p' and table_name ='th';
+------------+----------------+------------+----------------+-------------+
| table_name | partition_name | table_rows | avg_row_length | data_length |
+------------+----------------+------------+----------------+-------------+
| th         | p0             |          2 |             20 |          20 |
| th         | p1             |          0 |              0 |           0 |
+------------+----------------+------------+----------------+-------------+
2 rows in set (0.00 sec)
```

从结果可以看出 null 值被插入分区中。

9.10 本章练习

1. 举例说明范围分区和列表分区的不同以及二者的使用场景。
2. 创建一张用户表,表字段为用户编号、用户名称、出生年月,并设置出生年月为范围分区。
3. 修改用户表中出生年月的分区类型,并将年份作为列表分区的主要依据,例如 1990 年份出生的用户信息在同一个分区中。

第 10 章

MySQL 分区管理

本章内容：

- MySQL 分区的管理
- MySQL 分区维护
- MySQL 分区限制
- 分区键与主键、唯一键的关系

在 MySQL 中，随着数据的不断增加，分区管理变得越来越重要，本章将讲解 MySQL 分区的管理、维护、限制等内容，同时还会列举大量示例供读者理解和学习。

10.1 范围分区和列表分区的管理

使用 SQL 语句修改范围分区和列表分区时，可以使用 alter table 语句的分区扩展来添加、删除、重新定义、合并或拆分现有分区。

10.1.1 添加和删除分区

范围分区和列表分区的添加和删除操作相似。

1. 范围分区

下面是一张按范围分区的数据表：

给数据表分了 6 个分区

```
mysql> create table tr (id int, name varchar(50), purchased date)
    ->     partition by range( year(purchased) ) (
    ->         partition p0 values less than (1990),
    ->         partition p1 values less than (1995),
    ->         partition p2 values less than (2000),
    ->         partition p3 values less than (2005),
    ->         partition p4 values less than (2010),
    ->         partition p5 values less than (2015)
    ->     );
Query OK, 0 rows affected (0.28 sec)
# 把数据添加到数据表
mysql> insert into tr values
    ->     (1, 'desk organiser', '2003-10-15'),
    ->     (2, 'alarm clock', '1997-11-05'),
    ->     (3, 'chair', '2009-03-10'),
    ->     (4, 'bookcase', '1989-01-10'),
    ->     (5, 'exercise bike', '2014-05-09'),
    ->     (6, 'sofa', '1987-06-05'),
    ->     (7, 'espresso maker', '2011-11-22'),
    ->     (8, 'aquarium', '1992-08-04'),
    ->     (9, 'study desk', '2006-09-16'),
    ->     (10, 'lava lamp', '1998-12-25');
Query OK, 10 rows affected (0.05 sec)
Records: 10  Duplicates: 0  Warnings: 0
# 查看 P2 分区的数量
mysql> select * from tr partition (p2);
+------+-------------+------------+
| id   | name        | purchased  |
+------+-------------+------------+
|    2 | alarm clock | 1997-11-05 |
|   10 | lava lamp   | 1998-12-25 |
+------+-------------+------------+
2 rows in set (0.00 sec)
# 删除 P2 分区
mysql> alter table tr drop partition p2;
 Query OK, 0 rows affected (0.03 sec)
# 查看 p2 分区的数量
mysql> select * from tr partition (p2);
Empty set (0.00 sec)
```

如以上操作所示，当我们删除分区时，分区中的数据也会被删除。如果想要在不丢失数据的情况下更改表的分区，需要使用 alter table … add partition 语句。

例 10.1　范围分区的管理，SQL 语句如下：

```
# 创建一张有 3 个分区的数据表
 create table members (
    id int,
    fname varchar(25),
    lname varchar(25),
    dob date
)
partition by range( year(dob) ) (
    partition p0 values less than (1980),
    partition p1 values less than (1990),
```

```
        partition p2 values less than (2000)
);
# 给表members再增加一个分区，分区的范围值大于之前的年份，可以成功创建
alter table members add partition (partition p3 value less than (2021));

# 尝试给表members增加一个分区，分区的范围值小于之前分区的最小范围值
mysql> alter table members > add partition ( > partition n value less than (1970));
ERROR 1463 (HY000): value less than value must be strictly » increasing for each partition
```

当添加一个分区，其分区值小于之前的最小分区值时，将会抛出异常。我们需要重新组织分区来解决此问题，示例 SQL 语句及其执行结果如下：

```
alter table members
    reorganize partition p0 into (
        partition n0 values less than (1920),
        partition n1 values less than (1980)
);
# 查询表members的结果和分区
mysql> show create table members\G
*************************** 1. row ***************************
       Table: members
create Table: create table `members` (
  `id` int(11) default null,
  `fname` varchar(25) default null,
  `lname` varchar(25) default null,
  `dob` date default null
) engine=InnoDB default charset=latin1
/*!50100 partition by range ( year(dob))
(partition n0 value less than (1920) engine = InnoDB,
 partition n1 value less than (1980) engine = InnoDB,
 partition p1 value less than (1990) engine = InnoDB,
 partition p2 value less than (2000) engine = InnoDB,
 partition p3 value less than (2021) engine = InnoDB) */
1 row in set (0.00 sec)
```

从上面的结果可以看出，新增的分区已成功创建好了。

2. 列表分区

接下来看看列表分区是如何进行管理的。

例 10.2 列表分区的管理，示例 SQL 语句及其执行结果如下：

```
# 使用列表分区创建tt表
create table tt (
    id int,
    data int
)
partition by list(data) (
    partition p0 values in (5, 10, 15),
    partition p1 values in (6, 12, 18)
);
# 给表添加一个正常的新分区
alter table tt add partition (partition p2 values in (7, 14, 21));
# 尝试创建一个新分区，新分区包含之前分区的值
```

```
mysql> alter table tt add partition
    >   (partition np values in (4, 8, 12));
ERROR 1465 (HY000): Multiple definition of same constant »
                in list partitioning
```

由此可以看出，如果添加的新分区的值包含之前分区的值，则会抛出异常，创建分区失败。

10.1.2 拆分合并分区

除了添加分区和删除分区之外，还可以在保证数据不丢失的情况下，把一个分区拆分为多个分区或者把多个分区合并为一个分区。

例 10.3 在保证数据不丢失的情况下拆分分区与合并分区，示例 SQL 语句如下：

```
# 查询当前 members 表的结构和分区情况
mysql> show create table members\G
*************************** 1. row ***************************
       Table: members
Create Table: create table `members` (
  `id` int(11) default null,
  `fname` varchar(25) default null,
  `lname` varchar(25) default null,
  `dob` date default null
) engine=InnoDB default charset=latin1
/*!50100 partition by range ( year(dob))
(partition n0 value less than (1970) engine = InnoDB,
 partition n1 value less than (1980) engine = InnoDB,
 partition p1 value less than (1990) engine = InnoDB,
 partition p2 value less than (2000) engine = InnoDB,
 partition p3 value less than (2010) engine = InnoDB) */
1 row in set (0.00 sec)
# 把 n0 分区拆分成两个分区，分区名称分别是 s0, s1
alter table members reorganize partition n0 into (
    partition s0 values less than (1960),
    partition s1 values less than (1970)
);
# 把 s0、s1 分区合并成一个分区
alter table members reorganize partition s0,s1 into (
    partition p0 values less than (1970)
);
```

10.2 哈希分区和键分区的管理

哈希分区和键分区的添加与删除操作相似。

1. 哈希分区

例 10.4 哈希分区的管理，示例 SQL 语句及其执行结果如下：

```
# 创建一张具有 10 个哈希分区的数据表
```

```
create table clients (
    id int,
    fname varchar(30),
    lname varchar(30),
    signed date
)
partition by hash( month(signed) )
partitions 10;
# 把分区数量从 10 个变成 6 个
mysql> alter table clients coalesce partition 4;
Query OK, 0 rows affected (0.02 sec)
```

2. 键分区

例 10.5 键分区的管理,示例 SQL 语句及其执行结果如下:

```
# 创建一张具有 10 个键分区的表
mysql> create table clients_lk (
    ->     id int,
    ->     fname varchar(30),
    ->     lname varchar(30),
    ->     signed date
    -> )
    -> partition by linear key(signed)
    -> partitions 12;
Query OK, 0 rows affected (0.03 sec)
# 把键分区数量从 10 个变成 6 个
mysql> alter table clients_lk coalesce partition 4;
Query OK, 0 rows affected (0.06 sec)
Records: 0  Duplicates: 0  Warnings: 0
```

需要注意的是,coalesce partition 后面的数字表示要删除的分区数。因此,当删除的分区数大于总分区数时会报错,示例如下:

```
mysql> alter table clients coalesce partition 18;
ERROR 1478 (HY000): Cannot remove all partitions, use drop table instead
```

10.3　分区管理和维护操作

1. 删除分区(仅限于范围分区和列表分区,会丢失数据)

如果有多余的分区不想使用了,则可以通过 drop 指令删除分区。示例如下:

```
# 一次性删除一个分区
alter table emp drop partition p1;
# 一次性删除多个分区
alter table emp drop partition p1,p2;
```

2. 增加分区

如果要为现有的表增加分区,则可以通过 add 指令实现。示例如下:

```
# 增加范围分区
alter table emp add partition (partition p3 values less than (5000));
# 增加列表分区
alter table emp1 add partition (partition p3 values in (5000));
```

3. 分解分区（不会丢失数据）

reorganize partition 关键字可以对表的部分分区或全部分区进行修改，并且不会丢失数据。分解前后分区的整体范围应该一致。示例如下：

```
alter table t
reorganize partition p1 into
(
    partition p1 values less than (1000),
    partition p3 values less than (2000)
);
```

4. 合并分区（不会丢失数据）

随着分区数量的增多，有时需要把多个分区合并成一个分区，则可以使用 into 指令实现。示例如下：

```
alter table t
reorganize partition p1,p3 into
(partition p1 values less than (10000));
```

5. 重新定义哈希分区（不会丢失数据）

随着业务的不断扩展，想要对哈希分区进行扩容或缩容，则可以对现有的哈希分区进行重新定义。示例如下：

```
alter table t partition by hash(salary) partitions 8;
```

6. 重新定义范围分区（不会丢失数据）

随着业务的不断扩展，想要对范围分区进行扩容或缩容，则可以对现有的范围分区进行重新定义。示例如下：

```
alter table t partition by range(salary)
(
    partition p1 values less than (20000),
    partition p2 values less than (30000)
);
```

7. 删除表的所有分区（不会丢失数据）

在 MySQL 中，如果要删除表的所有分区，但又不想删除数据，那么可以执行如下语句：

```
alter table emp remove partitioning;
```

8. 重建分区

这和先删除保存在分区中的所有记录，然后重新插入它们具有同样的效果，可用于整理分区碎片。示例如下：

```
alter table emp rebuild partitionp1,p2;
```

9. 优化分区

如果从分区中删除了大量的行，或者对一个带有可变长度的行做了许多修改，那么可以使用 alter table … optimize partition 来收回没有使用的空间，并整理分区数据文件的碎片。示例如下：

```
alter table t optimize partition p1,p2;
```

10. 分析分区

在 MySQL 中，想要对现有的分区进行分析，可以执行如下语句：

```
# 读取并保存分区的键分布
alter table t analyze partition p1,p2;
```

11. 修补分区

在 MySQL 中，想要对现有的分区进行修补，可以执行如下语句：

```
# 修补被破坏的分区
alter table t repairpartition p1,p2;
```

12. 检查分区

在 MySQL 中，想要查看现有的分区是否被破坏，可以执行如下语句：

```
# 检查表指定的分区
alter table t check partition p1,p2;
```

这条 SQL 语句可以告诉我们表 t 的分区 P1、P2 中的数据或索引是否已经被破坏了。如果分区被破坏了，那么可以使用 alter table … repair partition 来修补该分区。

10.4 分区的限制

在业务中可以对分区进行一些限制：

（1）分区键必须包含在表的主键、唯一键中。

（2）MySQL 只能在使用分区函数的列进行比较时才能筛选分区，而不能根据表达式的值去筛选分区，即使这个表达式就是分区函数也不行。

（3）不使用 NDB 存储引擎的数据表的最大分区数为 8192。如果分区数很大，但是未达到 8192，则会出现错误提示信息 Got error … from storage engine: Out of resources when opening file，可以通过增加 open_files_limit 系统变量的值来解决这个问题，当然同时打开文件的数量也受操作系统本身的限制。

（4）InnoDB 存储引擎的分区不支持外键。

（5）服务器 SQL 模式（可以通过 SQL-MODE 参数进行配置）影响分区表的同步复制。主节点和从节点上不同的 SQL 模式可能会导致相同的数据存储在主从节点的不同分区中，甚至可能导致数据插入主节点成功，而插入从节点失败。为了获得最佳效果，应该始终在主机和从机上使用相同的服务器 SQL 模式，强烈建议不要在创建分区后更改服务器 SQL 模式。

（6）分区不支持全文索引，即使是使用 InnoDB 或 MyISAM 存储引擎的分区也不例外。

（7）分区无法使用外键约束。

（8）临时表不能进行分区。

10.5 分区键与主键、唯一键的关系

控制分区键与主键、唯一键关系的规则是：分区表达式中使用的所有列必须是该数据表可能具有的每个唯一键的一部分。换句话说，分区键必须包含在表的主键、唯一键中。

分区键与主键、唯一键关系的错误示例见例 10.6～例 10.9。

例 10.6 唯一键是 col1 和 col2，分区键是 col3，示例 SQL 语句及其执行结果如下：

```
mysql> create table t1 (
    ->     col1 int not null,
    ->     col2 date not null,
    ->     col3 int not null,
    ->     col4 int not null,
    ->     unique key (col1, col2)
    -> )
    -> partition by hash(col3)
    -> partitions 4;
ERROR 1503 (HY000): A PRIMARY KEY must include all columns in the table's partitioning function (prefixed columns are not considered).
```

因为 col3 不是 col1 和 col2 的一部分，所以报错。

例 10.7 两个唯一键分别是 col1 和 col3，分区键是根据运算符相加的 col1+col3，示例 SQL 语句及其执行结果如下：

```
mysql> create table t2 (
    ->     col1 int not null,
    ->     col2 date not null,
    ->     col3 int not null,
    ->     col4 int not null,
```

```
        -> unique key (col1),
        -> unique key (col3)
    -> )
    -> partition by hash(col1 + col3)
    -> partitions 4;
ERROR 1503 (HY000): A PRIMARY KEY must include all columns in the table's partitioning
function (prefixed columns are not considered).
```

因为分区键不是数据表的两个唯一键的一部分，所以报错。

例 10.8 两个唯一键分别是(col1, col2)和 col3，分区键为 col1 + col3，示例 SQL 语句及其执行结果如下：

```
mysql> create table t3 (
    ->   col1 int not null,
    ->   col2 date not null,
    ->   col3 int not null,
    ->   col4 int not null,
    ->   unique key (col1, col2),
    ->   unique key (col3)
    -> )
    -> partition by hash(col1 + col3)
    -> partitions 4;
ERROR 1491 (HY000): A PRIMARY KEY must include all columns in the table's partitioning
function
```

因为分区键 col1+col3 没有包含在两个唯一键中，所以报错。

例 10.9 主键是 col1 和 col2，分区键为 col3，示例 SQL 语句及其执行结果如下：

```
mysql> create table t5 (
    ->   col1 int not null,
    ->   col2 date not null,
    ->   col3 int not null,
    ->   col4 int not null,
    ->   primary key(col1, col2)
    -> )
    -> partition by hash(col3)
    -> partitions 4;
ERROR 1503 (HY000): A PRIMARY KEY must include all columns in the table's partitioning
function (prefixed columns are not considered).
```

因为分区键不在主键中，所以报错。

例 10.10 主键是 col1 和 col3，唯一键为 col2，分区键为 year(col2)，示例 SQL 语句及执行结果如下：

```
mysql> create table t6 (
    ->   col1 int not null,
    ->   col2 date not null,
    ->   col3 int not null,
    ->   col4 int not null,
    ->   primary key(col1, col3),
    ->   unique key(col2)
    -> )
```

```
        -> partition by hash( year(col2) )
        -> partitions 4;
ERROR 1503 (HY000): A PRIMARY KEY must include all columns in the table's partitioning
function (prefixed columns are not considered).
```

因为分区键没有包含在所有的唯一键中（包含主键），所以报错。

分区键与主键、唯一键关系的正确示例见例 10.10～例 10.13。

例 10.11 唯一键是复合键，包含 col1、col2 和 col3，分区键是 col3，示例 SQL 语句及其执行结果如下：

```
mysql> create table t1 (
    ->     col1 int not null,
    ->     col2 date not null,
    ->     col3 int not null,
    ->     col4 int not null,
    ->     unique key (col1, col2, col3)
    -> )
    -> partition by hash(col3)
    -> partitions 4;
Query OK, 0 rows affected (0.03 sec)
```

因为分区键是唯一键的一部分，所以执行正确。

例 10.12 唯一键是复合键，包含 col1 和 col3，分区键是 col1+col3，示例 SQL 语句及其执行结果如下：

```
mysql> create table t2 (
    ->     col1 int not null,
    ->     col2 date not null,
    ->     col3 int not null,
    ->     col4 int not null,
    ->     unique key (col1, col3)
    -> )
    -> partition by hash(col1 + col3)
    -> partitions 4;
Query OK, 0 rows affected (0.02 sec)
```

因为分区键包含在唯一键中，所以执行正确。

例 10.13 两个唯一键，分别为(col1, col2, col3)和 col3，分区键是 col3，示例 SQL 语句及其执行结果如下：

```
mysql> create table t3 (
    ->     col1 int not null,
    ->     col2 date not null,
    ->     col3 int not null,
    ->     col4 int not null,
    ->     unique key (col1, col2, col3),
    ->     unique key (col3)
    -> )
    -> partition by hash(col3)
    -> partitions 4;
```

```
Query OK, 0 rows affected (0.05 sec)
```

因为分区键包含在两个唯一键中,所以执行正确。

例 10.14 以下两种情况下,主键都不包括分区表达式中引用的所有列,但语句都是有效的。

情况 1:

```
create table t7 (
    col1 int not null,
    col2 date not null,
    col3 int not null,
    col4 int not null,
    primary key(col1, col2)
)
partition by hash(col1 + year(col2))
partitions 4;
```

情况 2:

```
create table t8 (
    col1 int not null,
    col2 date not null,
    col3 int not null,
    col4 int not null,
    primary key(col1, col2, col4),
    unique key(col2, col1)
)
partition by hash(col1 + year(col2))
partitions 4;
```

10.6 本章练习

1. 创建一张用户表,表字段为用户编号、用户名称、出生年月并将年份作为列表分区的主要依据,例如 1990 年份出生的用户信息在同一个分区。

2. 同上题,新增两个分区 p1 和 p2,分区年份分别为 2000 和 2010。

3. 同上题,合并 p1 和 p2 的分区为 p3。

4. 同上题,把 p3 还原为原来的 p1 和 p2 分区。

第 5 部分

视图、存储过程、字符集和排序规则的说明和使用

第 11 章　MySQL 视图和存储过程
第 12 章　MySQL 游标、字符集和排序规则

第 11 章

MySQL 视图和存储过程

本章内容：

- MySQL 视图
- MySQL 存储过程

视图是对表结构进行加密的最简单的方式，可以通过调用视图的方式查询数据、保存原始表的基础结构。存储过程是一种数据库对象，用于在数据库中存储复杂的程序以便外部程序调用。本章将详细讲解如何在 MySQL 中使用视图和存储过程，同时还会列举大量示例供读者理解和学习。

11.1 视　　图

视图是一个虚拟表，其内容是由查询语句来定义的。与真实的表一样，视图包含一系列带有名称的列数据和行数据。不过，视图并不在数据库中以存储数据集的形式存在，视图中的数据是根据定义视图的查询语句动态生成的。

视图是存储在数据库中的查询语句，它主要有以下两个优势：

- 视图可以隐藏一些数据，比如社会保险基金表可以用视图只显示姓名、地址，而不显示社会保险号和工资数等。
- 视图可以使复杂的查询更易于理解和使用。

视图就像一个"窗口"，从中只能看到想要的列数据（对应数据库中的列）和行数据（对应数据库中的记录）。这意味着可以视图上使用 select *，看到的将是在视图定义中给出的列数据和行数据。

视图带给我们了如下好处：

（1）视图能简化用户操作。视图机制使用户可以将注意力集中在所关心的数据上。如果这些数据不是直接来自数据库中的基本数据表，则可以通过定义视图使数据库看起来结构简单、清晰，并且可以简化用户的数据查询操作。例如，通过定义连接了多张表的视图，可以隐藏表与表之间的连接操作（对于数据库的用户而言）。换句话说，用户只需对一张虚拟表进行简单查询，而无需了解这张虚拟表是如何构建的。

（2）逻辑独立性。数据的物理独立性是指用户的应用程序不依赖于数据库的物理结构。数据的逻辑独立性是指当数据库重构（如增加新的关系或对原有的关系增加新的字段）时，用户的应用程序不会受影响。

（3）视图能够对机密数据提供安全保护。有了视图机制，就可以在设计数据库应用系统时对不同的用户定义不同的视图，使机密数据不出现在非授权用户的视图上。这样视图机制就自动提供了对机密数据的安全保护功能。例如，表 Student 涉及全校 15 个院系学生的数据，可以在其上定义 15 个视图，每个视图只包含一个院系的学生数据，并只允许每个院系的主任查询和修改本院系的学生视图。

例 11.1 使用视图，示例 SQL 语句及其执行结果如下：

```
# 创建表
mysql> create table t_employee(
    ->         id int  primary key  auto_increment,
    ->         name char (30) not null,
    ->         sex  char (2) not null,
    ->         age int not null,
    ->         department char (10) not null,
    ->         salary  int not null,
    ->         home char (30),
    ->         marry char (2) not null default '',
    ->         hobby char (30)
    ->  );
Query OK, 0 rows affected (0.02 sec)
# 添加数据
mysql> insert into t_employee(id, name , sex, age,department, salary, home, marry, hobby)
    -> values ( null , 'zhangsan' , '1' ,20, 'hr' , '4000' , 'guangndong' , 'no' , 'tennis' );
Query OK, 1 row affected (0.00 sec)

mysql> insert into t_employee(id, name , sex, age,department, salary, home, marry, hobby)
    -> values ( null , 'zhangsan2' , '1' ,22, 'hr' , '5000' , 'guangndong' , 'no' , 'tennis' );
Query OK, 1 row affected (0.01 sec)

mysql> insert into t_employee(id, name , sex, age,department, salary, home, marry, hobby)
```

```
    -> values ( null , 'zhangsan3' , '1' ,23, 'hr' , '6000' , 'guangndong' , 'no' , 'tennis' );
Query OK, 1 row affected (0.00 sec)
```

查询数据
```
mysql> select * from t_employee;
+----+-----------+-----+-----+------------+--------+-----------+-------+--------+
| id | name      | sex | age | department | salary | home      | marry | hobby  |
+----+-----------+-----+-----+------------+--------+-----------+-------+--------+
|  1 | zhangsan  |  1  |  20 | hr         |  4000  | guangndong| no    | tennis |
|  2 | zhangsan2 |  1  |  22 | hr         |  5000  | guangndong| no    | tennis |
|  3 | zhangsan3 |  1  |  23 | hr         |  6000  | guangndong| no    | tennis |
+----+-----------+-----+-----+------------+--------+-----------+-------+--------+
3 rows in set (0.00 sec)
```

创建视图
```
mysql> create view my_vie as select name,age,home from t_employee;
Query OK, 0 rows affected (0.01 sec)
```
查询视图的结果
```
mysql> select * from my_view;
+-----------+-----+------------+
| name      | age | home       |
+-----------+-----+------------+
| zhangsan  |  20 | guangndong |
| zhangsan2 |  22 | guangndong |
| zhangsan3 |  23 | guangndong |
+-----------+-----+------------+
3 rows in set (0.00 sec)
```

根据视图修改表中的数据，把年龄大于 20 岁的地址改为 xian
```
mysql> update my_view set home='xian' where age>20;
Query OK, 2 rows affected (0.01 sec)
Rows matched: 2  Changed: 2  Warnings: 0
```
查询视图数据
```
mysql> select * from my_view;
+-----------+-----+------------+
| name      | age | home       |
+-----------+-----+------------+
| zhangsan  |  20 | guangndong |
| zhangsan2 |  22 | xian       |
| zhangsan3 |  23 | xian       |
+-----------+-----+------------+
3 rows in set (0.00 sec)
```
查询原始表中的数据，发现原始表中的数据也被修改了
```
mysql> select id,name,sex,age,home from t_employee;
+----+-----------+-----+-----+------+
| id | name      | sex | age | home |
+----+-----------+-----+-----+------+
|  2 | zhangsan2 |  1  |  22 | xian |
|  3 | zhangsan3 |  1  |  23 | xian |
+----+-----------+-----+-----+------+
2 rows in set (0.00 sec)
```

根据视图删除数据 name='zhangsan'
```
mysql> delete from my_view where name='zhangsan';
```

```
Query OK, 1 row affected (0.00 sec)
# 验证 name='zhangsan'的数据是否被删除
mysql> select * from my_view;
+-----------+-----+------+
| name      | age | home |
+-----------+-----+------+
| zhangsan2 |  22 | xian |
| zhangsan3 |  23 | xian |
+-----------+-----+------+
2 rows in set (0.00 sec)
# 验证在原始表中 name='zhangsan'的数据是否也被删除
mysql> select id,name,sex,age,home from t_employee;
+----+-----------+-----+-----+------------+--------+------+
| id | name      | sex | age | department | salary | home |
+----+-----------+-----+-----+------------+--------+------+
|  2 | zhangsan2 |   1 |  22 | hr         |   5000 | xian |
|  3 | zhangsan3 |   1 |  23 | hr         |   6000 | xian |
+----+-----------+-----+-----+------------+--------+------+
2 rows in set (0.00 sec)
# 修改视图的语句
mysql> alter  view my_view as select id,name,sex,age,home from t_employee where age>20;
Query OK, 0 rows affected (0.00 sec)
# 验证视图是否已经修改成功
mysql> select * from my_view;
+----+-----------+-----+-----+------------+--------+------+
| id | name      | sex | age | department | salary | home |
+----+-----------+-----+-----+------------+--------+------+
|  2 | zhangsan2 |   1 |  22 | hr         |   5000 | xian |
|  3 | zhangsan3 |   1 |  23 | hr         |   6000 | xian |
+----+-----------+-----+-----+------------+--------+------+
2 rows in set (0.00 sec)
# 删除视图
mysql> drop view my_view ;
Query OK, 0 rows affected (0.01 sec)
# 验证视图是否已经删除成功
mysql> select * from my_view;
ERROR 1146 (42S02): Table 'demo.MY_VIEW' doesn't exist
```

11.2 存储过程

存储过程是一种数据库对象，用于在数据库中存储复杂的程序以便外部程序调用。存储过程是为了完成特定功能的一组 SQL 语句的集合，经过编译后保存在数据库中，用户可以通过存储过程的名字和给定的参数来调用与执行它们。存储过程的设计目标就是实现数据库 SQL 语言编写的程序的代码封装与代码复用（重复使用）。下面介绍使用存储过程的优点和缺点。

存储过程的优点如下：

（1）存储过程可以封装，并隐藏复杂的业务逻辑。

（2）存储过程可以回传值，还可以接收参数。

（3）存储过程无法使用 select 命令来运行，因为它是子程序，与查看表结构、表数据或用户定义函数不同。

存储过程的缺点如下：

（1）存储过程往往定制化于特定的数据库上，因为支持的编程语言不同，所以当切换到其他厂商的数据库系统时，需要重新编写原有的存储过程。

（2）当业务复杂时，存储过程的维护成本比较高。

11.2.1　存储过程的创建和调用

创建存储过程的语法如下：

```
create
    [definer = { user | current_user }]
procedure sp_name ([proc_parameter[,...]])
    [characteristic ...] routine_body

proc_parameter:
    [ in | out | inout ] param_name type

characteristic:
    comment 'string'
  | language sql
  | [not] deterministic
  | { contains sql | no sql | reads sql data | modifies sql data }
  | sql security { definer | invoker }

routine_body:
    valid sql routine statement

[begin_label:] begin
    [statement_list]
       ……
end [end_label]
```

例 11.2　创建和使用存储过程，示例 SQL 语句及其执行结果如下：

```
# 数据准备、创建表与新增数据
mysql> create table userinfo (
    -> id int,
    -> name varchar (100),
    -> age int
    -> );
Query OK, 0 rows affected (0.03 sec)
mysql> insert into userinfo values(1,'zhangsan',18);
Query OK, 1 row affected (0.00 sec)
mysql> insert into userinfo values(2,'zhangsan2',19);
Query OK, 1 row affected (0.01 sec)
```

```
mysql> insert into userinfo values(3,'zhangsan3',20);
Query OK, 1 row affected (0.01 sec)

# 创建一个根据 id 删除用户信息的存储过程
# 临时修改结束符号，防止创建存储过程的语句不完整执行
mysql> delimiter $
mysql> create procedure proc_delete_userinfo(in p_id integer)
    -> begin
    -> delete from userinfo where id = p_id;
    -> end$
Query OK, 0 rows affected (0.01 sec)

# 调用存储过程，删除 id=1 的用户，然后验证是否删除成功
# 修改结束符号
mysql> delimiter;
mysql> select * from userinfo;
+------+-----------+------+
| id   | name      | age  |
+------+-----------+------+
|   1  | zhangsan  |  18  |
|   2  | zhangsan2 |  19  |
|   3  | zhangsan3 |  20  |
+------+-----------+------+
3 rows in set (0.00 sec)
# 调用存储过程
mysql> call proc_delete_userinfo(1);
Query OK, 1 row affected (0.00 sec)
# 查询结果，验证 id=1 的用户信息是否已经被删除
mysql> select * from userinfo;
+------+-----------+------+
| id   | name      | age  |
+------+-----------+------+
|   2  | zhangsan2 |  19  |
|   3  | zhangsan3 |  20  |
+------+-----------+------+
2 rows in set (0.00 sec)
```

11.2.2 带参数的存储过程

MySQL 存储过程的参数用于存储过程的定义，共有 in、out 和 inout 三种参数类型，其语法如下：

```
create procedure 存储过程名([[in |out |inout ] 参数名 数据类型...])
```

参数类型说明：

- in（输入参数）：表示调用者向过程传入值。
- out（输出参数）：表示过程向调用者传出值。
- inout（输入输出参数）：既表示调用者向过程传入值，又表示过程向调用者传出值。

1. in（输入）参数

例 11.3　输入参数的存储过程，示例 SQL 语句及其执行结果如下：

```
mysql> delimiter $$
mysql> create procedure in_param(in p_in int)
    -> begin
    ->     select p_in;
    ->     set p_in=2;
    ->     select P_in;
    -> end$$
mysql> delimiter;
# 定义变量
mysql> set @p_in=1;
mysql> call in_param(@p_in);
+------+
| p_in |
+------+
|    1 |
+------+

+------+
| P_in |
+------+
|    2 |
+------+
# 查询变量的值
mysql> select @p_in;
+-------+
| @p_in |
+-------+
|     1 |
+-------+
```

从本例可以看出，p_in 在存储过程中被修改，但并不影响@p_in 的值，因为前者为局部变量，后者为全局变量。

2. out（输出）参数

例 11.4　输出参数的存储过程，示例 SQL 语句及其执行结果如下：

```
# 设置结束符为$
mysql> delimiter $
# 创建存储过程
mysql> create procedure proc_delete_userinfo2(in p_id integer,out msg varchar(100))
    -> begin
    ->     delete from userinfo where id = p_id;
    ->     set msg='delete is ok';
    -> end$
Query OK, 0 rows affected (0.00 sec)
# 设置结束符
mysql> delimiter;
# 查询表数据
mysql> select * from userinfo;
+------+-----------+------+
| id   | name      | age  |
+------+-----------+------+
|    1 | zhangsan  |   18 |
|    2 | zhangsan2 |   19 |
```

```
|   3 | zhangsan3 |   20 |
+------+-----------+------+
3 rows in set (0.00 sec)
# 设置输入变量
mysql> set @id=1;
Query OK, 0 rows affected (0.00 sec)
# 设置输出变量
mysql> set @msg='';
Query OK, 0 rows affected (0.00 sec)
# 调用存储过程
mysql> call proc_delete_userinfo2(@id,@msg);
Query OK, 1 row affected (0.00 sec)
# 查询输出变量的值
mysql> select @msg;
+--------------+
| @msg         |
+--------------+
| delete is ok |
+--------------+
1 row in set (0.00 sec)
# 验证数据是否被删除
mysql> select * from userinfo;
+------+-----------+------+
| id   | name      | age  |
+------+-----------+------+
|   2  | zhangsan2 |   19 |
|   3  | zhangsan3 |   20 |
+------+-----------+------+
2 rows in set (0.00 sec)
```

3. inout（输入输出）参数

例 11.5 输入输出参数的存储过程，示例 SQL 语句及其执行结果如下：

```
# 设置结束符为$
mysql> delimiter $$
# 创建存储过程
mysql> create procedure inout_param(inout p_inout int)
    -> begin
    ->   select p_inout;
    ->   set p_inout=2;
    ->   select p_inout;
    -> end
    -> $$
# 设置结束符
mysql> delimiter;
# 设置输入输出变量
mysql> set @p_inout=1;
# 调用存储过程
mysql> call inout_param(@p_inout);
+---------+
| p_inout |
+---------+
|       1 |
+---------+
```

```
+---------+
| p_inout |
+---------+
|       2 |
+---------+
# 查询输入输出变量的值
mysql> select @p_inout;
+----------+
| @p_inout |
+----------+
|        2 |
+----------+
```

11.2.3 变量

在存储过程中，可以使用变量来存储和处理数据，从而使存储过程更加灵活和可重用。在 MySQL 存储过程中，变量可以分为四类：用户变量、局部变量、会话级变量和全局变量。

- 用户变量：是在 SQL 语句执行时由数据库系统自动创建和管理的变量。它们通常用于存储函数或存储过程返回的结果或计算结果。用户变量的作用范围是会话级别，即只在当前会话内有效。
- 局部变量：是在 SQL 语句执行时由程序员手动创建和管理的变量。它们通常用于存储函数或存储过程内部的临时数据。局部变量的作用范围是当前会话级别，即只在当前会话内有效。
- 会话级变量：是在会话开始时由数据库系统自动创建和管理的变量。它们通常用于存储全局数据或共享数据。会话级变量的作用范围是整个会话，即在整个会话期间有效。
- 全局变量：由系统提供，整个 MySQL 服务器有效。

1. 变量定义

局部变量的声明一定要放在存储过程的开始，语法如下：

```
declare variable_name [,variable_name...] datatype [default value];
```

例 11.6 局部变量的声明，示例 SQL 语句如下：

```
declare l_numeric number(8,2) default 10.25;
declare l_date date default '2019-12-31';
declare l_varchar varchar(255) default 'this is proc';
```

2. 变量赋值

变量赋值的语法如下：

```
set 变量名 = 表达式值 [,variable_name = expression ...]
```

3. 用户变量

用户变量是会话级别的变量，其变量的作用域仅限于声明它的客户端连接。当这个客户端断开时，其所有的会话变量都将被释放。

例 11.7 在 MySQL 客户端使用用户变量，示例 SQL 语句及其执行结果如下：

```
mysql > select 'hello world' into @x;
mysql > select @x;
+-------------+
| @x          |
+-------------+
| hello world |
+-------------+
mysql > set @z=1+2+3+4+5;
mysql > select @z;
+------+
| @z   |
+------+
| 15   |
+------+
# 在存储过程中使用用户变量
mysql > create procedure proc_world( ) select concat(@a,' world');
mysql > set @a='hello';
mysql > call proc_world( );
+--------------------------+
| concat(@a,' world')      |
+--------------------------+
| Hello World              |
+--------------------------+
# 在存储过程之间传递全局范围的用户变量
mysql> create procedure p1()    set @last_procedure='p1';
mysql> create procedure p2() select concat('last procedure was ', @last_procedure);
mysql> call p1( );
mysql> call p2( );
+----------------------------------------------+
| concat('last procedure was ',@last_proc      |
+----------------------------------------------+
| Last procedure was p1                        |
+----------------------------------------------+
```

11.2.4 注释

MySQL 存储过程可以使用两个横杠（--）或"/**/"来进行注释。

例 11.8 在存储过程中使用两个横杠来注释，示例 SQL 语句如下：

```
# 将 MySQL 的分隔符以默认的分号改为//，以便在存储过程中可以使用分号
mysql > delimiter //
mysql > create procedure proc1 -- name 存储过程名（这是注释）
    -> (in parameter1 integer)
    -> begin
    -> declare variable1 char(10);
    -> if parameter1 = 17 then
    -> set variable1 = 'birds';
    -> else
    -> set variable1 = 'beasts';
```

```
    -> end if;
    -> insert into table1 values (variable1);
    -> end
    -> //
```

11.2.5 存储过程的修改和删除

修改存储过程的语法如下：

alter procedure 存储过程名称

删除存储过程的语法如下：

 drop procedure 存储过程名称

11.2.6 存储过程的控制语句

1. 变量作用域

内部变量在其作用域范围内享有更高的优先权，当执行到 end 关键字时，内部变量将从作用域中消失，因为它已经超出了其作用域范围。此时，该变量不再可见了，因为在存储过程外部再也找不到这个变量的声明，但是，我们可以通过 out 参数或者赋值方式来保存会话变量的值。

例 11.9 变量作用域的示例 SQL 语句及其执行结果如下：

```
mysql > delimiter //
mysql > create procedure proc3()
    -> begin
    -> declare x1 varchar(3) default "aaa";
    -> begin
    -> declare x1 varchar(3) default "bbb";
     -> select x1;
     -> end;
      -> select x1;
    -> end;
    -> //
mysql> call proc3();

+------+
| x1   |
+------+
| bbb  |
+------+
1 row in set (0.00 sec)

+------+
| x1   |
+------+
| aaa  |
+------+
1 row in set (0.00 sec)
```

这是一个 MySQL 存储过程，其中包含两个 begin...end 块。在第一个块中，声明了一个名为 x1 的字符串变量并赋值为"aaa"。接下来，进入第二个块，在其中再次声明一个名为 x1 的字符串变量并赋值为"bbb"，然后在第二个块的 select 语句中调用 x1。接着在第二个块的外部、第一个块的末尾再一次在 select 语句中调用 x1。最后，我们通过 call proc3() 来执行存储过程，并打印出 x1 的值。执行结果显示在一个表格中，其中只有一行，它的 x1 值为"bbb"。

这是由于在 MySQL 中，变量作用域仅限于它们声明的块内部。如果在嵌套块中声明同名的变量，那么内部块中声明的变量将覆盖外部块中声明的同名变量，即第二个 begin...end 块内的 x1 变量是当前的活动变量，该块内声明的 x1 变量覆盖了外部块的 x1 变量。因此，当选择该变量的值时，其值是"bbb"。执行完内部块后，内部块的 x1 变量被销毁，外部块中声明的 x1 变量才会成为当前的活动变量，即在选择第二个 x1 变量的值时，其值是"aaa"。

2. 条件语句

条件语句可以给定一个判断条件，并在程序执行过程中判断该条件是否成立，根据判断结果执行不同的操作，从而改变代码的执行顺序，以实现更多的功能。

1）if-then-else 语句

if-then-else 的逻辑看似简单，功能却十分强大，它可以根据不同的条件执行不同的操作。

例 11.10　if-then-else 语句的示例 SQL 语句如下：

```
mysql > delimiter //
mysql > create procedure proc2(in parameter int)
    -> begin
    -> declare var int;
    -> set var=parameter+1;
    -> if var=0 then
    -> insert into t values(18);
    -> end if;
    -> if parameter=0 then
    -> update t set s1=s1+80;
    -> else
    -> update t set s1=s1+12;
    -> end if;
    -> end;
    -> //
```

以上 SQL 代码创建了一个名为 proc2 的存储过程，该存储过程采用一个整数类型的参数 parameter。在存储过程的代码块中，首先声明了一个名为 var 的整型变量，并将它初始化为 parameter+1。然后，我们使用一个条件语句来检查 var 是否等于 0，如果 var 等于 0，则插入一个值为 18 的行到表 t 中；如果 var 不等于 0，则继续使用条件语句判断参数 parameter 的值是否等于 0，若 parameter 等于 0，则更新表 t 中的 s1 列的值为 s1+80，否则更新 s1 列的值为 s1+12。

2）case 语句

在 MySQL 中，除了使用 if-then-else 语句外，还可以使用 case 语句来根据不同的条件执行不同的操作，case 语句相对于 if-then-else 更简洁。

例 11.11　case 语句的示例 SQL 语句如下：

```
mysql > delimiter //
mysql > create procedure proc3 (in parameter int)
    -> begin
    -> declare var int;
    -> set var=parameter+1;
    -> case var
    -> when 0 then
    -> insert into t values(17);
    -> when 1 then
    -> insert into t values(18);
    -> else
    -> insert into t values(19);
    -> end case;
    -> end;
    -> //
```

以上 SQL 代码创建了一个名为 proc2 的存储过程，该存储过程采用一个整数类型的参数 parameter。在存储过程的代码块中，首先声明了一个名为 var 的整型变量，并将它初始化为 parameter+1；然后使用一个条件语句来检查 var 的值，并根据 var 不同的值进行不同的操作：如果 var 等于 0，则向 t 表中插入一行 17 的数据；如果 var 等于 1，则向 t 表中插入一行 18 的数据；如果 var 的值不等于 0 也不等于 1，则向 t 表中插入一行 19 的数据。

3. 循环语句

在实际开发中，会有许多具有规律的重复操作，因此在程序中就需要重复执行某些语句，控制这些语句反复执行的语句就是循环语句。

1）while 语句

while 语句是 MySQL 中的一种基本循环语句。当满足条件时进入循环体，当不满足条件时跳出循环体。

例 11.12　while 循环语句的示例 SQL 语句如下：

```
mysql > delimiter $
mysql > create procedure proc4()
    -> begin
    -> declare var int;
    -> set var=0;
    -> while var<6 do
    -> insert into t values(var);
    -> set var=var+1;
    -> end while;
    -> end;
```

```
->$
```

以上 SQL 代码创建了一个名为 proc4 的存储过程，该存储过程包含一个 while 循环。在存储过程的代码块中，首先声明了一个名为 var 的整型变量，并将其初始值设置为 0。然后使用 while 循环，如果 var 的值小于 6，则插入一个值为 var 的行到表 t 中，并将 var 的值自增 1。重复执行，直到 var 的值达到 6，结束 while 循环。

2）loop 语句

在 MySQL 中，除了使用 while 语句外，还可以使用 loop 语句来处理重复的操作。在实际开发中，用户可以根据自己的喜好来选择。这两种语句除了语法不太一样外，实际效果是相同的。

例 11.13 loop 循环语句的示例 SQL 语句如下：

```
mysql > delimiter $
mysql > create procedure proc6 ()
    -> begin
    -> declare v int;
    -> set v=0;
    -> loop_lable:loop
    -> insert into t values(v);
    -> set v=v+1;
    -> if v >=5 then
    -> leave loop_lable;
    -> end if;
    -> end loop;
    -> end;
    -> //
```

以上 SQL 代码创建了一个名为 proc6 的存储过程，该存储过程包含一个 loop 循环。在代码块中，首先定义一个名为 v 的整型变量，并将它初始化为 0；然后使用 loop 来执行一个无限循环，每次都将变量 v 插入表 t 中，并将变量 v 的值加 1；接着使用 if 语句检查变量 v 的值是否大于或等于 5，如果是，则使用 leave 退出循环。

11.3 本章练习

1. 写出创建视图的语法。
2. 写出创建存储过程的语法。
3. 说明视图的使用场景和优点。
4. 说明存储过程的使用场景和优点。

第 12 章

MySQL 游标、字符集和排序规则

本章内容：

- MySQL 游标
- MySQL 字符集和排序规则

本章将讲解如何在 MySQL 中使用游标，以及 MySQL 中支持的字符集和字符集的排序规则，同时还会列举大量示例供读者理解和学习。

12.1 游 标

MySQL 在服务器端提供只读的、单向的游标，而且只能在存储过程或者更底层的客户端 API 中使用。因为 MySQL 游标中指向的对象都存储在临时表中，而不是指向数据库中对应的实际数据，所以 MySQL 游标总是只读的。

1. 游标的特性

- 不敏感：数据库可以选择不复制结果集。
- 只读。
- 不滚动和回滚：游标只能向同一方向前进，并且不可以跳过中间任何一行的数据。

2. 游标的优点

游标是针对行操作的，对从数据库中查询得到的结果集的每一行可以分开、独立地执行相同或者不相同的操作。

3. 游标的缺点

- 性能不高，只能一行一行地操作。
- 使用游标可能会产生死锁，造成内存开销过大。

4．游标的适用场景

- 存储过程。
- 函数。
- 触发器。

例 12.1 使用游标的示例 SQL 语句及其执行结果如下：

```
# 准备数据
create table student(
    stuId int primary key auto_increment,
    stuName varchar(20),
    stuSex varchar(2),
    stuAge int
)default charset=utf8;
# 创建一个普通游标，查询年龄大于 18 岁的学生的信息
delimiter //
create procedure p1()
begin
    declare id int;
    declare name varchar(100) character set utf8;
    declare done int default 0;
    -- 声明游标
    declare mc cursor for select stuId,stuName from student where stuAge >18;
    declare continue handler for not found set done = 1;
    -- 打开游标
    open mc;
    -- 获取结果
    fetch mc into id,name;
    -- 这里是为了显示获取的结果
    select id,name;
    -- 关闭游标
    close mc;
end //
delimiter ;

# 结合 loop 创建游标
 -- 定义语法结束符号
delimiter //
-- 创建一个名称为 p2 的存储过程
create procedure p2()
begin
    -- 声明用于接收游标指向的变量
    declare id,age,total int;
    -- 注意，接收游标指向的值为中文时，需要给变量指定字符集为 utf8
    declare name,sex varchar(20) character set utf8;
    -- 游标结束的标志
    declare done int default 0;
    -- 声明游标
    declare cur cursor for select stuId,stuName,stuSex,stuAge from student where stuAge > 18;
    -- 指定游标循环结束时的返回值
    declare continue handler for not found set done =1;
    -- 打开游标
```

```sql
    open cur;
    -- 初始化变量
    set total = 0;
    -- loop 循环
    u:loop
        -- 根据游标当前指向的一条数据
        fetch cur into id,name,sex,age;
        -- 当游标的返回值为 1 时,退出 loop 循环
        if done = 1 then
            leave u;
        end if;
        -- 累计
        set total = total + 1;
    end loop;
    -- 关闭游标
    close cur;
    -- 输出累计的结果
    select total;
end //
delimiter ;

# 结合 while 创建游标
delimiter //
-- 创建一个名称为 p3 的存储过程
create procedure p3()
begin
    -- 声明用于接收游标指向值的变量
    declare id,age,total int;
    -- 注意,接收游标指向的值为中文时,需要给变量指定字符集为 utf8
    declare name,sex varchar(20) character set utf8;
    -- 游标结束的标志
    declare done int default 0;
    -- 声明游标
    declare cur cursor for select stuId,stuName,stuSex,stuAge from student where stuAge > 18;
    -- 指定游标循环结束时的返回值
    declare continue handler for not found set done = 1;
    -- 打开游标
    open cur;
    -- 初始化变量
    set total = 0;
    -- while 循环
    while done != 1 do
        fetch cur into id,name,sex,age;
        if done != 1 then
            set total = total + 1;
        end if;
    end while;
    -- 关闭游标
    close cur;
    -- 输出累计的结果
    select total;
end //
delimiter ;
```

12.2　字符集和排序规则

MySQL 支持多种字符集，使我们能够使用各种字符集来存储数据并根据排序规则进行字符或字符串的比较。MySQL 服务器默认的字符集和排序规则是 utf8mb4 和 utf8mb4_0900_ai_ci，不过我们可以根据具体情况为服务器、数据库、表和列指定字符集。字符集问题不仅会影响数据的存储，还会影响客户端程序和 MySQL 服务器之间的通信。如果希望客户端程序不使用默认的字符集与服务器通信，那么需要指明具体的字符集。例如，要使用 utf8 Unicode 字符集，则在连接到服务器后执行以下语句：

```
set names 'utf8';
```

12.2.1　MySQL 中的字符集和排序规则

1. 字符集

MySQL 服务器支持多种字符集，包括多个 Unicode 字符集。可以使用 show character set 语句查询当前服务器支持的字符集。

例 12.2　使用 show character set 语句查询当前服务器支持的字符集，SQL 语句及其执行结果如下：

```
mysql> show character set;
+----------+-----------------------------+---------------------+--------+
|Charset   | Description                 | Default collation   |Maxlen  |
+----------+-----------------------------+---------------------+--------+
| big5     | Big5 Traditional Chinese    | big5_chinese_ci     |     2  |
| dec8     | DEC West European           | dec8_swedish_ci     |     1  |
| cp850    | DOS West European           | cp850_general_ci    |     1  |
| hp8      | HP West European            | hp8_english_ci      |     1  |
| koi8r    | KOI8-R Relcom Russian       | koi8r_general_ci    |     1  |
| latin1   | cp1252 West European        | latin1_swedish_ci   |     1  |
| latin2   | ISO 8859-2 Central European | latin2_general_ci   |     1  |
| swe7     | 7bit Swedish                | swe7_swedish_ci     |     1  |
| ascii    | US ASCII                    | ascii_general_ci    |     1  |
| ujis     | EUC-JP Japanese             | ujis_japanese_ci    |     3  |
| sjis     | Shift-JIS Japanese          | sjis_japanese_ci    |     2  |
| hebrew   | ISO 8859-8 Hebrew           | hebrew_general_ci   |     1  |
| tis620   | TIS620 Thai                 | tis620_thai_ci      |     1  |
| euckr    | EUC-KR Korean               | euckr_korean_ci     |     2  |
| koi8u    | KOI8-U Ukrainian            | koi8u_general_ci    |     1  |
| gb2312   | GB2312 Simplified Chinese   | gb2312_chinese_ci   |     2  |
| greek    | ISO 8859-7 Greek            | greek_general_ci    |     1  |
| cp1250   | Windows Central European    | cp1250_general_ci   |     1  |
| gbk      | GBK Simplified Chinese      | gbk_chinese_ci      |     2  |
| latin5   | ISO 8859-9 Turkish          | latin5_turkish_ci   |     1  |
| armscii8 | ARMSCII-8 Armenian          | armscii8_general_ci |     1  |
| utf8     | UTF-8 Unicode               | utf8_general_ci     |     3  |
| ucs2     | UCS-2 Unicode               | ucs2_general_ci     |     2  |
| cp866    | DOS Russian                 | cp866_general_ci    |     1  |
```

```
| keybcs2    | DOS Kamenicky Czech-Slovak      | keybcs2_general_ci    |   1 |
| macce      | Mac Central European            | macce_general_ci      |   1 |
| macroman   | Mac West European               | macroman_general_ci   |   1 |
| cp852      | DOS Central European            | cp852_general_ci      |   1 |
| latin7     | ISO 8859-13 Baltic              | latin7_general_ci     |   1 |
| utf8mb4    | UTF-8 Unicode                   | utf8mb4_general_ci    |   4 |
| cp1251     | Windows Cyrillic                | cp1251_general_ci     |   1 |
| utf16      | UTF-16 Unicode                  | utf16_general_ci      |   4 |
| utf16le    | UTF-16LE Unicode                | utf16le_general_ci    |   4 |
| cp1256     | Windows Arabic                  | cp1256_general_ci     |   1 |
| cp1257     | Windows Baltic                  | cp1257_general_ci     |   1 |
| utf32      | UTF-32 Unicode                  | utf32_general_ci      |   4 |
| binary     | Binary pseudo charset           | binary                |   1 |
| geostd8    | GEOSTD8 Georgian                | geostd8_general_ci    |   1 |
| cp932      | SJIS for Windows Japanese       | cp932_japanese_ci     |   2 |
| eucjpms    | UJIS for Windows Japanese       | eucjpms_japanese_ci   |   3 |
| gb18030    | China National Standard GB18030 | gb18030_chinese_ci    |   4 |
+------------+---------------------------------+-----------------------+-----+
41 rows in set (0.00 sec)
```

默认情况下，show character set 语句会显示出所有可用的字符集。我们可以使用 like 或者 where 子句来指示要匹配的字符集名称。

例 12.3 使用 like 子句匹配字符集名称 utf%，示例 SQL 语句及其执行结果如下：

```
mysql> show character set like 'utf%';
+---------+-----------------+--------------------+--------+
| Charset | Description     | Default collation  | Maxlen |
+---------+-----------------+--------------------+--------+
| utf8    | UTF-8 Unicode   | utf8_general_ci    |   3    |
| utf8mb4 | UTF-8 Unicode   | utf8mb4_general_ci |   4    |
| utf16   | UTF-16 Unicode  | utf16_general_ci   |   4    |
| utf16le | UTF-16LE Unicode| utf16le_general_ci |   4    |
| utf32   | UTF-32 Unicode  | utf32_general_ci   |   4    |
+---------+-----------------+--------------------+--------+
5 rows in set (0.00 sec)
```

2. 排序规则

一个给定的字符集至少有一个支持的排序规则，大多数字符集有几个支持的排序规则。

例 12.4 查看默认字符集 utf8mb4 的排序规则，示例 SQL 语句及其执行结果如下：

```
mysql> show collation where charset = 'utf8mb4';
+------------------------+---------+-----+---------+----------+---------+
| Collation              | Charset | Id  | Default | Compiled | Sortlen |
+------------------------+---------+-----+---------+----------+---------+
| utf8mb4_general_ci     | utf8mb4 |  45 | Yes     | Yes      |    1    |
| utf8mb4_bin            | utf8mb4 |  46 |         | Yes      |    1    |
| utf8mb4_unicode_ci     | utf8mb4 | 224 |         | Yes      |    8    |
| utf8mb4_icelandic_ci   | utf8mb4 | 225 |         | Yes      |    8    |
| utf8mb4_latvian_ci     | utf8mb4 | 226 |         | Yes      |    8    |
| utf8mb4_romanian_ci    | utf8mb4 | 227 |         | Yes      |    8    |
| utf8mb4_slovenian_ci   | utf8mb4 | 228 |         | Yes      |    8    |
| utf8mb4_polish_ci      | utf8mb4 | 229 |         | Yes      |    8    |
```

```
| utf8mb4_estonian_ci    | utf8mb4 | 230 |     | Yes     |       8 |
| utf8mb4_spanish_ci     | utf8mb4 | 231 |     | Yes     |       8 |
| utf8mb4_swedish_ci     | utf8mb4 | 232 |     | Yes     |       8 |
| utf8mb4_turkish_ci     | utf8mb4 | 233 |     | Yes     |       8 |
| utf8mb4_czech_ci       | utf8mb4 | 234 |     | Yes     |       8 |
| utf8mb4_danish_ci      | utf8mb4 | 235 |     | Yes     |       8 |
| utf8mb4_lithuanian_ci  | utf8mb4 | 236 |     | Yes     |       8 |
| utf8mb4_slovak_ci      | utf8mb4 | 237 |     | Yes     |       8 |
| utf8mb4_spanish2_ci    | utf8mb4 | 238 |     | Yes     |       8 |
| utf8mb4_roman_ci       | utf8mb4 | 239 |     | Yes     |       8 |
| utf8mb4_persian_ci     | utf8mb4 | 240 |     | Yes     |       8 |
| utf8mb4_esperanto_ci   | utf8mb4 | 241 |     | Yes     |       8 |
| utf8mb4_hungarian_ci   | utf8mb4 | 242 |     | Yes     |       8 |
| utf8mb4_sinhala_ci     | utf8mb4 | 243 |     | Yes     |       8 |
| utf8mb4_german2_ci     | utf8mb4 | 244 |     | Yes     |       8 |
| utf8mb4_croatian_ci    | utf8mb4 | 245 |     | Yes     |       8 |
| utf8mb4_unicode_520_ci | utf8mb4 | 246 |     | Yes     |       8 |
| utf8mb4_vietnamese_ci  | utf8mb4 | 247 |     | Yes     |       8 |
+------------------------+---------+-----+---------+---------+---------+
26 rows in set (0.00 sec)
```

需要注意的是，排序规则具有以下特征：

（1）两个不同的字符集不能有相同的排序规则。

（2）每个字符集都有一个默认的排序规则。

12.2.2 列字符集和排序规则

每个字符列（类型为 char、varchar、text 或任何同义词的列）都有一个列字符集和一个列排序规则，用于列定义的 create table 和 alter table 语句具有用于指定列的字符集和归类的可选子句，子句语法如下：

```
col_name {char | varchar | text} (col_length)
    [character set charset_name]
    [collate collation_name]
```

例 12.5　创建数据表时就指定好列字符集和排序规则，示例 SQL 语句如下：

```
create table t1
(
    col1 varchar(5)
      character set latin1
      collate latin1_german1_ci
);
```

例 12.6　使用 alter 的方式修改数据表中的列字符集和排序规则，示例 SQL 语句如下：

```
alter table t1 modify
    col1 varchar(5)
    character set latin1
    collate latin1_swedish_ci;
```

12.2.3 使用 collate 查询排序规则

使用 collate 子句可以覆盖用于比较的任何默认排序规则,当多个列使用相同的字符集时,它们将按照 collate 子句进行排序。collate 可用于 SQL 语句的各个部分,示例如下。

例 12.7 collate 与 order by 结合查询排序规则,示例 SQL 语句如下:

```
select k
from t1
order by k collate latin1_german2_ci;
```

例 12.8 collate 与 as 结合查询排序规则,示例 SQL 语句如下:

```
select k collate latin1_german2_ci as k1
from t1
order by k1;
```

例 12.9 collate 与 group by 结合查询排序规则,示例 SQL 语句如下:

```
select k
from t1
group by k collate latin1_german2_ci;
```

例 12.10 collate 与聚合函数结合查询排序规则,示例 SQL 语句如下:

```
select max(k collate latin1_german2_ci)
from t1;
```

例 12.11 collate 与 distinct 结合查询排序规则,示例 SQL 语句如下:

```
select distinct k collate latin1_german2_ci
from t1;
```

例 12.12 collate 与 where 结合查询排序规则,示例 SQL 语句如下:

```
select *
from t1
where _latin1 'müller' collate latin1_german2_ci = k;
elect *
from t1
where k like _latin1 'müller' collate latin1_german2_ci;
```

例 12.13 collate 与 having 结合查询排序规则,示例 SQL 语句如下:

```
select k
from t1
group by k
having k = _latin1 'müller' collate latin1_german2_ci;
```

12.3 本章练习

1. 写出创建游标的语法。
2. 说明游标的使用场景和优点。

第 6 部分　MySQL 数据查询优化

第 13 章　MySQL 索引
第 14 章　InnoDB 存储引擎事务模型和锁使用
第 15 章　InnoDB 内存结构解析和优化

第 13 章

MySQL 索引

本章内容：

- MySQL 索引介绍
- MySQL 索引的指导原则
- B-Tree 索引和哈希索引的比较
- SQL 执行计划分析
- MySQL 索引的检索原理
- MySQL 索引的使用技巧

本章将讲解 MySQL 的索引的基本知识，各种索引的创建和使用。如果想要提升 SQL 语句的性能，那么本章尤为重要，本章包含索引的底层和检索原理，这是编写 SQL 语句必须掌握的技能。

13.1 索　引

索引（Index）是可以帮助 MySQL 高效获取数据的数据结构。索引的目的在于提高查询效率，可以将它与字典的目录类比，例如要查 mysql 这个单词，我们肯定要先定位到字母 m，然后从上往下查找字母 y，再查找剩下的 sql。如果没有索引，我们可能需要扫描整部字典，这会导致查询性能下降，所以索引是一种提升查询性能、快速高效地获取数据的方法。

数据库除了存储业务数据之外，还维护着一个满足特定查找算法的数据结构，这个数据结构以某种方式指向数据，这样就可以在数据结构的基础上实现高级查找算法，这种数据结构就是索引。一般来说，索引本身也很大，不可能全部存储在内存中，因此往往以索引文件的形式存储在磁盘上。

常用的索引有如下几类：

- 普通索引：基本索引类型，允许在定义索引的列中插入重复值和空值。
- 唯一索引：索引列数据不允许出现重复，但是允许有空值。
- 主键索引：要求主键列中的每个值是非空、唯一的。创建一个主键后，系统将会自动创建主键索引。
- 复合索引：将多个列组合作为索引。
- 前缀索引：只能作用于文本内容，也就是只能在 text、char 和 varchar 数据类型的列上创建前缀索引，可以指定索引列的长度。

下面是索引的基本操作。

（1）创建索引，[unique]可以省略：

```
create [unique] index indexname on tabname(column1,column2);
```

只写一个字段就是单值索引，写多个字段就是复合索引。

（2）删除索引：

```
drop index [indexname] on tabname;
```

（3）查看索引：

```
show index from tabname \g;
```

（4）通过修改数据表的方式创建索引：

```
# 该语句表示给数据表添加一个主键，系统会默认给此主键字段创建主键索引。这意味着索引值必须是唯一的，并且不能为null，需要注意的是多个字段也可以作为一个主键
alter table tabname add primary key(column_list);

# 该语句表示创建一个唯一索引，索引的值必须是唯一的（除了null之外）
alter table tabname add unique indexname(column_list);

# 该语句表示创建一个普通索引，索引值可以重复
alter table tabname add index indexname(column_list);

# 该语句表示创建一个前缀索引
alter table tabname add index(column(4));

# 该语句表示创建了一个fulltext索引，主要用于全文检索
alter table tabname add fulltext indexname(column_list);
```

操作示例如下：

例 13.1 创建索引并对索引进行操作，示例 SQL 语句及其执行结果如下：

```
# 创建一张用户表
mysql> create table userinfo(
    -> id int not null,
    -> name varchar(100),
```

```
    -> address varchar(255),
    -> age int,
    -> cardid varchar(20)
    -> );
Query OK, 0 rows affected (0.00 sec)
# 创建 id 为主键索引
mysql> alter table userinfo add primary key(id);
Query OK, 0 rows affected (0.00 sec)
Records: 0  Duplicates: 0  Warnings: 0
# 创建 cardid 为唯一索引
mysql> alter table userinfo add unique key inx_uq_cardid(cardid);
Query OK, 0 rows affected (0.00 sec)
Records: 0  Duplicates: 0  Warnings: 0

# 创建 address 为普通索引
mysql> alter table userinfo add index inx_address(address);
Query OK, 0 rows affected (0.01 sec)
Records: 0  Duplicates: 0  Warnings: 0
# 创建 name 和 address 为复合索引
mysql> alter table userinfo add index inx_name_address(name,address);
Query OK, 0 rows affected (0.01 sec)
Records: 0  Duplicates: 0  Warnings: 0
# 查询当前用户表中所有索引的详细情况
mysql> show index from userinfo \G;
*************************** 1. row ***************************
        Table: userinfo
   Non_unique: 0
     Key_name: PRIMARY
 Seq_in_index: 1
  Column_name: id
    Collation: A
  Cardinality: 0
     Sub_part: NULL
       Packed: NULL
         Null:
   Index_type: BTREE
      Comment:
Index_comment:
*************************** 2. row ***************************
        Table: userinfo
   Non_unique: 0
     Key_name: inx_uq_cardid
 Seq_in_index: 1
  Column_name: cardid
    Collation: A
  Cardinality: 0
     Sub_part: NULL
       Packed: NULL
         Null: YES
   Index_type: BTREE
      Comment:
Index_comment:
*************************** 3. row ***************************
        Table: userinfo
   Non_unique: 1
     Key_name: inx_address
```

```
   Seq_in_index: 1
   Column_name: address
      Collation: A
    Cardinality: 0
       Sub_part: NULL
         Packed: NULL
           Null: YES
     Index_type: BTREE
        Comment:
  Index_comment:
*************************** 4. row ***************************
          Table: userinfo
     Non_unique: 1
       Key_name: inx_name_address
   Seq_in_index: 1
    Column_name: name
      Collation: A
    Cardinality: 0
       Sub_part: NULL
         Packed: NULL
           Null: YES
     Index_type: BTREE
        Comment:
  Index_comment:
*************************** 5. row ***************************
          Table: userinfo
     Non_unique: 1
       Key_name: inx_name_address
   Seq_in_index: 2
    Column_name: address
      Collation: A
    Cardinality: 0
       Sub_part: NULL
         Packed: NULL
           Null: YES
     Index_type: BTREE
        Comment:
  Index_comment:
5 rows in set (0.00 sec)
```

13.2 创建索引的指导原则

MySQL 在以下操作中使用索引：
- 快速查找与 where 子句匹配的行。
- 在多个索引之间进行选择时，MySQL 通常选择使用查找行数最少的索引。
- 如果数据表具有多列索引，则优化器可以使用索引的任何最左前缀来查找数据。例如，如果在（col1, col2, col3）上有一个 3 列索引，则在使用(col1)、(col1, col2)和(col1, col2, col3)的查询功能时都能使用索引（最左匹配原则）。
- 如果需要进行类型转换才能比较值，则不同数据类型的列进行比较（例如，将字符串列与数字列进行比较）时可能会阻止使用索引。
- 如果查询仅使用某个索引中包含的列，则可以直接从索引树中检索所选值，以提高查询效率（索引覆盖）。

> **提　示**
> 当查询需要访问表中的大多数行时，那么按顺序读取比使用索引更快，因为顺序读取可以最大限度地减少磁盘搜索。

13.3　B-Tree 索引和哈希索引的比较

MySQL 支持的索引数据结构有两种：一种是 B-Tree（默认的索引存储结构），另一种是哈希。

1. B-Tree 索引

图 13-1 是 B-Tree 索引的示例图。

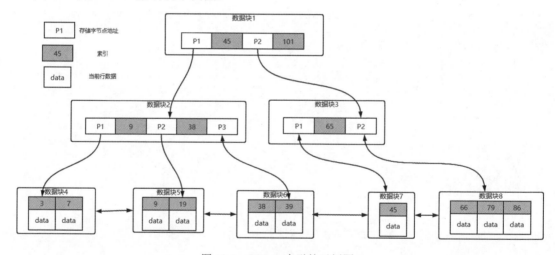

图 13-1　B-Tree 索引的示例图

B-Tree 索引具有如下特性：

- 所有关键字都出现在叶子节点的链表中，并且链表中的关键字恰好是有序的。
- 非叶子节点相当于叶子节点的索引，叶子节点相当于存储关键字数据的数据层。非根节点存储的其实是指向根节点的索引。
- 数据值存储在树的末梢（即存储在叶子节点中）。
- 根节点横向也有链指针（方便范围查询、分页查询）。

通过上面几点可以发现 B-Tree 类型适合在某一叶子节点到另一叶子节点范围内进行查找，例如用于<、>、orderby 等应用场景中。

2. 哈希索引

哈希索引是一种检索效率非常高的精确定位索引。哈希索引不像 B-Tree 索引那样需要从根节点到子节点进行多次访问才能访问到叶子节点,因此哈希索引的查询效率要远高于 B-Tree 索引,它会将计算出的哈希值和对应的行指针信息记录在哈希表中。虽然哈希索引效率很高,但是同样有很多弊端和限制。哈希索引的特征如下:

- 哈希索引仅能满足"="、"in"和"<=>"查询,不能使用范围查询。
- 哈希索引无法避免数据的排序操作。
- 哈希索引不能利用部分索引键查询。
- 哈希索引在任何时候都不能避免表的扫描查找(遇到重复哈希值的情况)。
- 只能使用整个键来搜索一行。

13.4 SQL 执行计划分析

在日常工作中,有时会使用慢查询去记录一些执行时间比较长的 SQL 语句,找出这些 SQL 语句后,使用 explain 命令来分析这些 SQL 语句的执行计划,查看该 SQL 语句有没有使用索引,有没有做全表扫描。explain 命令可以帮助我们深入了解 MySQL 基于开销的优化器,还可以获得很多可能被优化器考虑到的访问策略的细节,以及当运行 SQL 语句时哪种策略预计会被优化器采用。因此,使用 explain 命令来了解执行计划是我们优化 SQL 必备的手段。explain 命令可以解决以下问题:

(1)数据表的读取顺序是什么。
(2)数据读取操作有哪些操作类型。
(3)哪些索引可以被使用。
(4)哪些索引被实际使用。
(5)数据表之间如何引用。
(6)每张数据表有多少行被优化器查询。

接下来将通过示例来深入地理解和学习 explain 命令。

例 13.2 使用 explain 命令来了解执行计划。

首先进行数据准备,示例 SQL 语句如下:

```
# 创建学科数据表
create table subject(
  id int(10) auto_increment,
  name varchar(20),
  teacher_id int(10),
  primary key (id),
  index idx_teacher_id (teacher_id));
```

```
# 创建教师表
create table teacher(
  id int(10) auto_increment,
  name varchar(20),
  teacher_no varchar(20),
  primary key (id),
  unique index unx_teacher_no (teacher_no(20)));

# 创建学生表
create table student(
  id int(10) auto_increment,
  name varchar(20),
  student_no varchar(20),
  primary key (id),
  unique index unx_student_no (student_no(20)));

# 创建学生成绩表
create table student_score(
  id int(10) auto_increment,
  student_id int(10),
  subject_id int(10),
  score int(10),
  primary key (id),
  index idx_student_id (student_id),
  index idx_subject_id (subject_id));
# 在教师表中增加 name 字段为普通索引
alter table teacher add index idx_name(name(20));
# 数据填充
 insert into student(name,student_no) values
('zhangsan','20200001'),('lisi','20200002'),('yan','20200003'),('dede',
'20200004');
 insert into teacher(name,teacher_no) values
('wangsi','T2010001'),('sunsi','T2010002'),('jiangsi','T2010003'),
('zhousi','T2010004');
 insert into subject(name,teacher_id) values
('math',1),('Chinese',2),('English',3),('history',4);
insert into student_score(student_id,subject_id,score) values
(1,1,90),(1,2,60),(1,3,80),(1,4,100),(2,4,60),(2,3,50),(2,2,80),(2,1,90),
(3,1,90),(3,4,100),(4,1,40),(4,2,80),(4,3,80),(4,5,100);
```

接着使用 explain 命令分析 select * from subject 语句，看看能得到什么结果，示例 SQL 语句及其执行结果如下：

```
mysql> explain select * from subject \G;
*************************** 1. row ***************************
           id: 1
  select_type: SIMPLE
        table: subject
   partitions: NULL
         type: ALL
possible_keys: NULL
          key: NULL
      key_len: NULL
          ref: NULL
         rows: 4
```

```
        filtered: 100.00
           Extra: NULL
1 row in set, 1 warning (0.00 sec)
```

从上面的结构可以看出，explain 输出了很多列，每一列中字段的意义如下：

- id：表示 select 查询的序列号，包含一组数字，也就是查询中执行 select 子句或操作表的顺序。
- select_type：表示查询类型。
- table：表示正在访问哪张数据表。
- partitions：表示匹配的分区。
- type：表示访问的类型。
- possible_keys：表示可能应用在这张表中的一个或多个索引，但不一定会实际用到。
- key：表示实际用到的索引，如果为 null，则没有使用索引。
- key_len：表示索引中使用的字节数，可以通过该列计算查询中使用的索引的长度。
- ref：表示显示索引的哪一列被使用了，一般是一个常数。
- rows：表示根据数据表统计信息及索引的选用情况大致估算出所需记录的读取行数。
- filtered：表示查询的表行占数据表的百分比。
- Extra：表示不适合在其他列中显示但又十分重要的额外信息。

1. id 列

（1）如果 id 的值都相同，则表示执行顺序是从上到下；如果 id 的值不同，则 id 的值越大优先级越高，越先被执行。

例 13.3　出现多行 id，且 id 值相同，示例 SQL 语句及其执行结果如下：

```
mysql> explain select subject.* from subject,student_score,teacher where subject.id
= student_id and subject.teacher_id = teacher.id \G;
*************************** 1. row ***************************
           id: 1
  select_type: SIMPLE
        table: subject
   partitions: NULL
         type: ALL
possible_keys: PRIMARY,idx_teacher_id
          key: NULL
      key_len: NULL
          ref: NULL
         rows: 4
     filtered: 100.00
        Extra: Using where
*************************** 2. row ***************************
           id: 1
  select_type: SIMPLE
        table: teacher
   partitions: NULL
         type: eq_ref
possible_keys: PRIMARY
```

```
        key: PRIMARY
    key_len: 4
        ref: demo.subject.teacher_id
       rows: 1
   filtered: 100.00
      Extra: Using index
*************************** 3. row ***************************
         id: 1
select_type: SIMPLE
      table: student_score
 partitions: NULL
       type: ref
possible_keys: idx_student_id
        key: idx_student_id
    key_len: 5
        ref: demo.subject.id
       rows: 1
   filtered: 100.00
      Extra: Using index
3 rows in set, 1 warning (0.00 sec)
```

通过上面的返回结果可以看出，3张表的id值都是1，查询数据库时，将按顺序从上往下执行，查询表的顺序是 subject > teacher > student_score。

（2）如果出现多行id，其值有的相同，有的不同，则可以认为id相同的是一组，在同一组中从上往下顺序执行；不同的组id值越大，优先级越高，越先执行。

例 13.4 出现多行id，且id值不相同，示例SQL语句及其执行结果如下：

```
mysql> explain select subject.* from subject left join teacher on subject.teacher_id
 = teacher.id
    -> union
    -> select subject.* from subject right join teacher on subject.teacher_id =
teacher.id \G;
*************************** 1. row ***************************
         id: 1
select_type: PRIMARY
      table: subject
 partitions: NULL
       type: ALL
possible_keys: NULL
        key: NULL
    key_len: NULL
        ref: NULL
       rows: 4
   filtered: 100.00
      Extra: NULL
*************************** 2. row ***************************
         id: 1
select_type: PRIMARY
      table: teacher
 partitions: NULL
       type: eq_ref
possible_keys: PRIMARY
```

```
            key: PRIMARY
        key_len: 4
            ref: demo.subject.teacher_id
           rows: 1
       filtered: 100.00
          Extra: Using index
*************************** 3. row ***************************
             id: 2
    select_type: UNION
          table: teacher
     partitions: NULL
           type: index
  possible_keys: NULL
            key: unx_teacher_no
        key_len: 23
            ref: NULL
           rows: 4
       filtered: 100.00
          Extra: Using index
*************************** 4. row ***************************
             id: 2
    select_type: UNION
          table: subject
     partitions: NULL
           type: ALL
  possible_keys: idx_teacher_id
            key: NULL
        key_len: NULL
            ref: NULL
           rows: 4
       filtered: 100.00
          Extra: Using where; Using join buffer (Block Nested Loop)
*************************** 5. row ***************************
             id: NULL
    select_type: UNION RESULT
          table: <union1,2>
     partitions: NULL
           type: ALL
  possible_keys: NULL
            key: NULL
        key_len: NULL
            ref: NULL
           rows: NULL
       filtered: NULL
          Extra: Using temporary
5 rows in set, 1 warning (0.00 sec)
```

从上面的结果可以看出，id 值依次是 subject（primary）=1、teacher（primary）=1、teacher（union）=2、subject（union）=2，所以表的查询顺序是 teacher（union）>subject（union）>subject（primary）>teacher（primary）。

2. select_type 字段

select_type 字段有 10 种取值，每种值表示的含义如表 13-1 所示。

表 13-1　select_type 的每种值表示的含义

select_type 值	含　义
simple	表示简单 select（不使用 union 或子查询）
primary	表示最外层查询被标记为主查询
union	表示若第二个 select 出现在 union 之后，则被标记为 union
dependent union	表示 select 语句中的第二个或后面的 union 会被标记为此类型
union result	表示 union 的结果集
dependent subquery	表示 select 在子查询中，依赖于外部查询
derived	表示在 from 列表中包含的子查询被标记为 derived（派生）
dependent derived	表示派生表依赖于另一张表
materialized	表示物化子查询
uncacheable subquery	表示子查询无法缓存结果并且必须为外部查询的每一行重新评估

（1）select_type=simple，表示此 SQL 语句是简单查询，不包含子查询或 union 查询。

例 13.5　select_type=simple 的示例 SQL 语句及其执行结果如下：

```
mysql> explain select subject.* from subject,student_score,teacher where subject.id
= student_id and subject.teacher_id = teacher.id \G;
*************************** 1. row ***************************
           id: 1
  select_type: SIMPLE
        table: teacher
   partitions: NULL
         type: index
possible_keys: PRIMARY
          key: unx_teacher_no
      key_len: 23
          ref: NULL
         rows: 4
     filtered: 100.00
        Extra: Using index
*************************** 2. row ***************************
           id: 1
  select_type: SIMPLE
        table: subject
   partitions: NULL
         type: ALL
possible_keys: PRIMARY,idx_teacher_id
          key: NULL
      key_len: NULL
          ref: NULL
         rows: 4
     filtered: 25.00
        Extra: Using where; Using join buffer (Block Nested Loop)
*************************** 3. row ***************************
           id: 1
  select_type: SIMPLE
        table: student_score
   partitions: NULL
```

```
        type: ref
possible_keys: idx_student_id
         key: idx_student_id
     key_len: 5
         ref: demo.subject.id
        rows: 1
    filtered: 100.00
       Extra: Using index
3 rows in set, 1 warning (0.00 sec)
```

从上面的结果可以看出，3 行数据的 select_type 都是 simple，说明这 3 张表都进行了简单查询。

（2）select_type=primary，表示查询中包含任何复杂的子部分，最外层查询被标记为主查询。

例 13.6 select_type=primary 的示例 SQL 语句及其执行结果如下：

```
mysql> explain select score.* from student_score as score where subject_id = (select id from subject where teacher_id = (select id from teacher where id = 2)) \G;
*************************** 1. row ***************************
          id: 1
 select_type: PRIMARY
       table: student_score
  partitions: NULL
        type: ref
possible_keys: idx_subject_id
         key: idx_subject_id
     key_len: 5
         ref: const
        rows: 3
    filtered: 100.00
       Extra: Using where
*************************** 2. row ***************************
          id: 2
 select_type: SUBQUERY
       table: subject
  partitions: NULL
        type: ref
possible_keys: idx_teacher_id
         key: idx_teacher_id
     key_len: 5
         ref: const
        rows: 1
    filtered: 100.00
       Extra: Using where; Using index
*************************** 3. row ***************************
          id: 3
 select_type: SUBQUERY
       table: teacher
  partitions: NULL
        type: const
possible_keys: PRIMARY
         key: PRIMARY
     key_len: 4
         ref: const
```

```
        rows: 1
    filtered: 100.00
       Extra: Using index
3 rows in set, 1 warning (0.01 sec)
```

通过上面的结果可以看出,student-score 表的 select_type 等于 primary,因此上述 SQL 语句中 student_score 表被作为主表进行查询。

3. type 字段

type 字段有 10 种取值,每种值表示的含义如表 13-2 所示。

表 13-2 每种 type 值表示的含义

type 值	含 义
null	表示 MySQL 能够在优化阶段分解查询语句,在执行阶段不用再访问表或索引
system	表示数据表只有一行记录(等于系统表),这是 const 类型的特例,一般的 SQL 语句出现的概率很小
const	表示通过索引一次就找到了结果,const 用于比较 primary key 或 unique 索引,因为只匹配一行数据,所以很快,如主键置于 where 列表中,MySQL 就能将该查询转换为一个常量
eq_ref	表示匹配的分区
ref	表示非唯一性索引扫描,返回匹配某个单独值的所有行
ref_or_null	类似于 ref,但是可以搜索值为 null 的行
index_merge	表示使用了索引合并的优化方法
range	表示只检索给定范围的数据行,使用一个索引来选择行,key 列显示使用了哪个索引。一般就是在 where 子句中出现 between、<>、in 等的查询
index	表示查询只遍历索引树,通常比使用 all 快
all	表示将遍历全数据表以找到匹配的行

以上几种值的查询性能由快到慢依次为:

null>system>const>eq_ref>ref>ref_or_null>index_merge>range>index>all;

(1)type=null,表示 MySQL 能够在优化阶段分解查询语句,在执行阶段不用访问数据表或通过索引就可以查询出结果。

例 13.7 type=null 的示例 SQL 语句及其执行结果如下:

```
mysql> explain select min(id) from subject \G;
*************************** 1. row ***************************
           id: 1
  select_type: SIMPLE
        table: NULL
   partitions: NULL
         type: NULL
possible_keys: NULL
          key: NULL
      key_len: NULL
          ref: NULL
         rows: NULL
     filtered: NULL
```

```
        Extra: Select tables optimized away
1 row in set, 1 warning (0.00 sec)
```

从上面的结果可以看出，查询最小 id 的时候，type=null，说明此语句被优化，不需要查询表或索引就可以查询出结果。

（2）type=all，表示此 SQL 语句的所有数据都是从硬盘上扫描整个数据表获得的，这种方式的性能是最差的。

例 13.8　type=all 的示例 SQL 语句及其执行结果如下：

```
mysql> explain select * from subject \G;
*************************** 1. row ***************************
           id: 1
  select_type: SIMPLE
        table: subject
   partitions: NULL
         type: ALL
possible_keys: NULL
          key: NULL
      key_len: NULL
          ref: NULL
         rows: 4
     filtered: 100.00
        Extra: NULL
1 row in set, 1 warning (0.00 sec)
```

从上面的结果可以看出，查询 subject 表时没有使用任何 where 子句来筛选，因此没有用到任何索引，此语句的结果是通过在硬盘上进行全表扫描获得的，在业务开发中应尽量避免采用这种方式查询。

（3）type=index，表示此 SQL 语句先用索引获得了数据指针，再去查询原始数据，或者直接从索引中得到结果。一般情况下，这种方式比全表扫描速度更快，性能更高。

例 13.9　type=index 的示例 SQL 语句及其执行结果如下：

```
mysql> explain select id from subject \G;
*************************** 1. row ***************************
           id: 1
  select_type: SIMPLE
        table: subject
   partitions: NULL
         type: index
possible_keys: NULL
          key: idx_teacher_id
      key_len: 5
          ref: NULL
         rows: 4
     filtered: 100.00
        Extra: Using index
1 row in set, 1 warning (0.00 sec)
```

从上面的结果可以看出，type=index 表示查询了索引，尽管这条 SQL 语句没有使用 where

等子句，但是查询的字段是 id，而 id 是此表的主键，默认是聚集索引，因此经过优化器分析，这些数据可以全部来源于索引数据，不需要在原始表中查询，也就是说不需要扫描索引。

（4）type=ref，表示非唯一性索引扫描，返回匹配某个单独值的所有行。type=ref 本质上也是一种索引访问，返回所有匹配某个单独值的行。这种扫描可能会找到多个符合条件的行，此扫描方式属于查找和扫描的混合体。

例 13.10 type=ref 的示例 SQL 语句及其执行结果如下：

```
mysql> explain select * from subject where teacher_id=9 \G;
*************************** 1. row ***************************
           id: 1
  select_type: SIMPLE
        table: subject
   partitions: NULL
         type: ref
possible_keys: idx_teacher_id
          key: idx_teacher_id
      key_len: 5
          ref: const
         rows: 1
     filtered: 100.00
        Extra: NULL
1 row in set, 1 warning (0.00 sec)
```

4. table 字段

此字段用于查询数据来源于哪张表。

例 13.11 table 字段的示例 SQL 语句及其执行结果如下：

```
mysql> explain select id from subject  \G;
*************************** 1. row ***************************
           id: 1
  select_type: SIMPLE
        table: subject
   partitions: NULL
         type: index
possible_keys: NULL
          key: idx_teacher_id
      key_len: 5
          ref: NULL
         rows: 4
     filtered: 100.00
        Extra: Using index
1 row in set, 1 warning (0.00 sec)
```

通过 table 字段的值就能发现此语句会去查询哪张表。

5. possible_keys 字段

此字段表示可能应用在这张表中的索引有一个或多个。查询涉及的字段若存在索引，则该索引将被列出，但不一定被实际使用。

例 13.12 possible_keys 字段的示例 SQL 语句及其执行结果如下：

```
mysql> explain select * from subject where teacher_id=9 \G;
*************************** 1. row ***************************
           id: 1
  select_type: SIMPLE
        table: subject
   partitions: NULL
         type: ref
possible_keys: idx_teacher_id
          key: idx_teacher_id
      key_len: 5
          ref: const
         rows: 1
     filtered: 100.00
        Extra: NULL
1 row in set, 1 warning (0.00 sec)
```

可以看出此语句可能用到了 idx_teacher_id 索引，但只是猜测。

6. key 字段

表示实际使用到的索引，如果为 null，则没有使用索引。查询中若使用了覆盖索引（查询的列刚好是索引），则该索引仅出现在 key 列表中。

例 13.13 key 字段的示例 SQL 语句及其执行结果如下：

```
mysql> explain select * from subject where teacher_id=9 \G;
*************************** 1. row ***************************
           id: 1
  select_type: SIMPLE
        table: subject
   partitions: NULL
         type: ref
possible_keys: idx_teacher_id
          key: idx_teacher_id
      key_len: 5
          ref: const
         rows: 1
     filtered: 100.00
        Extra: NULL
1 row in set, 1 warning (0.00 sec)
```

由结果可以看出，此语句用到了索引 idx_teacher_id。

7. key_len 字段

表示索引中使用的字节数，可通过该列计算查询中使用的索引的长度。在不损失精确度的情况下，索引的长度越短越好。key_len 显示的值为索引字段最大的可能长度，并非实际使用长度，即 key_len 是根据定义计算而得到的，不是表内检索出的。

例 13.14 key_len 字段的示例 SQL 语句及其执行结果如下：

```
mysql> explain select * from teacher where name='sunsi' \G;
*************************** 1. row ***************************
           id: 1
  select_type: SIMPLE
        table: teacher
   partitions: NULL
         type: ref
possible_keys: idx_name
          key: idx_name
      key_len: 63
          ref: const
         rows: 1
     filtered: 100.00
        Extra: NULL
1 row in set, 1 warning (0.00 sec)
```

可以看出此 SQL 查询用到了索引 idx_name，而且使用索引的长度是 63 字节。下面介绍影响索引长度的因素和索引使用字节数的计算方式。

（1）影响索引长度的因素：

- 列长度。
- 列是否为空，如 null(+1)、not null(+0)。
- 字符集，如 utf8mb4=4、utf8=3、gbk=2、latin1=1。
- 列类型为字符，如 varchar(+2)、char(+0)。
- int 占 4 字节。
- bigint 占 8 字节。

（2）计算公式：

key_len=（表字符集长度）×列长度+1（null）+2（变长列）

计算示例如下：

① charact_set=utf8, char(50), null

　　key_len=（3×50+1+0）=151

② charact_set=utf8, char(50), not null

　　key_len=（3×50+0+0）=150

③ charact_set=utf8, varchar(50), null

　　key_len=（3×50+1+2）=153

8. ref 字段

表示哪些列或常量将与 key 列中显示的索引进行比较，以从数据表中选择行。

例 13.15 ref 字段的示例 SQL 语句及其执行结果如下：

```
mysql> explain select * from teacher where name='sunsi' \G;
*************************** 1. row ***************************
           id: 1
  select_type: SIMPLE
        table: teacher
   partitions: NULL
         type: ref
possible_keys: idx_name
          key: idx_name
      key_len: 63
          ref: const
         rows: 1
     filtered: 100.00
        Extra: NULL
1 row in set, 1 warning (0.00 sec)
```

如上所示，ref 是一个常量，表示语句通过索引 idx_name 与常量进行比较，使用 idx_name 值从数据表中选择数据行。

9. row 字段

表示根据数据表统计信息及索引选用情况大致估算所需的记录需要读取的行数。

例 13.16 row 字段的示例 SQL 语句及其执行结果如下：

```
mysql> explain select * from teacher \G;
*************************** 1. row ***************************
           id: 1
  select_type: SIMPLE
        table: teacher
   partitions: NULL
         type: ALL
possible_keys: NULL
          key: NULL
      key_len: NULL
          ref: NULL
         rows: 4
     filtered: 100.00
        Extra: NULL
1 row in set, 1 warning (0.00 sec)
```

此语句表示进行全表扫描，扫描数量是 4 行，而此数据表的数量也刚好是 4 行。在优化 SQL 的时候，数据条数越少，查询速度越快。

10. partitions 字段

表示查询的数据从哪些分区获取。如果使用了分区，则尽量让此字段有值。这样就只需要查询数据表中的一部分数据，可以提高查询效率。

例 13.17 partitions 字段的示例 SQL 语句及其执行结果如下：

```
# 创建表 hash_part，对字段 id 以哈希方式分为 3 个分区
create table if not exists `hash_part` (
```

```
  `id` int(11) not null ,
  `comment` varchar(1000) not null default '' ,
  `ip` varchar(25) not null default '' ,
  primary key (`id`)
) engine=innodb default charset=utf8 auto_increment=1
partition by hash(id)
partitions 3;
# 插入数据
insert into hash_part values(1,'aaa','192.168.3.2');
insert into hash_part values(2,'bbb','192.168.3.2');
insert into hash_part values(3,'ccc','192.168.3.2');
insert into hash_part values(4,'ddd','192.168.3.2');
# 没有where子句的查询
mysql> explain select * from hash_part \G;
*************************** 1. row ***************************
           id: 1
  select_type: SIMPLE
        table: hash_part
   partitions: p0,p1,p2
         type: ALL
possible_keys: NULL
          key: NULL
      key_len: NULL
          ref: NULL
         rows: 4
     filtered: 100.00
        Extra: NULL
1 row in set, 1 warning (0.00 sec)
```

如上所示，partitions 字段为 p0,p1,p2，说明扫描了所有的分区。

```
# 根据分区的id来筛选
mysql> explain select * from hash_part where id=1 \G;
*************************** 1. row ***************************
           id: 1
  select_type: SIMPLE
        table: hash_part
   partitions: p1
         type: const
possible_keys: PRIMARY
          key: PRIMARY
      key_len: 4
          ref: const
         rows: 1
     filtered: 100.00
        Extra: NULL
1 row in set, 1 warning (0.00 sec)
```

如上所示，partitions 字段为 p1，说明只从 p1 分区中查询了数据。

> **提　示**
>
> 如果对数据表进行了分区，那么在查询时让 partitions 字段中的分区数量尽可能地小，这样就能提高查询效率。

11. extra 字段

extra 字段包含不适合在其他列中显示但又十分重要的一些额外信息。extra 值及其含义如表 13-3 所示。

表 13-3 extra 值及其含义

extra 值	含 义
using filesort	表示 MySQL 会对数据使用一个外部的索引排序,而不是按照表内的索引顺序进行读取
using temporary	表示使用了临时表保存中间结果,MySQL 在对结果排序时使用临时表,常见于排序（order by）和分组查询（group by）
using index	表示相应的 select 操作中使用了覆盖索引,避免访问表的数据行,效率比较高。如果同时出现 using where,则表明索引被用来执行索引键值的查找。如果没有同时出现 using where,则表明索引用来读取数据而非执行查找动作
using where	表示使用了 where 子句
using join buffer	表示使用了连接缓存
impossible where	表示 where 子句的值总是 false,不能用来获取任何数据
distinct	表示一旦 MySQL 找到了相匹配的数据行,就不再进行搜索
select tables optimized away	表示 MySQL 根本没有遍历表或索引就返回数据了,select 操作已经优化到不能再优化了

（1）extra=using filesort，表示 MySQL 会对数据使用一个外部的索引排序，而不是按照表内的索引顺序进行读取。

例 13.18　extra=using filesort 的示例 SQL 语句及其执行结果如下：

```
mysql> explain select * from subject order by name \G;
*************************** 1. row ***************************
           id: 1
  select_type: SIMPLE
        table: subject
   partitions: NULL
         type: ALL
possible_keys: NULL
          key: NULL
      key_len: NULL
          ref: NULL
         rows: 4
     filtered: 100.00
        Extra: Using filesort
1 row in set, 1 warning (0.00 sec)
```

如上查询 subject 数据需要排序输出。这个排序字段不是索引字段（如果是索引字段，则需要判断 key 字段的值是否和排序字段一致），因此使用了外部排序。

（2）extra=distinct，表示一旦 MySQL 找到了与数据表中行列值相匹配的行数据，就不再继续搜索了。

例 13.19　extra= distinct 的示例 SQL 语句及其执行结果如下：

```
mysql> explain select distinct teacher.name from teacher left join subject on
teacher.id = subject.teacher_id \G;
*************************** 1. row ***************************
           id: 1
  select_type: SIMPLE
        table: teacher
   partitions: NULL
         type: index
possible_keys: idx_name
          key: idx_name
      key_len: 63
          ref: NULL
         rows: 4
     filtered: 100.00
        Extra: Using index; Using temporary
*************************** 2. row ***************************
           id: 1
  select_type: SIMPLE
        table: subject
   partitions: NULL
         type: ref
possible_keys: idx_teacher_id
          key: idx_teacher_id
      key_len: 5
          ref: demo.teacher.id
         rows: 1
     filtered: 100.00
        Extra: Using index; Distinct
2 rows in set, 1 warning (0.00 sec)
```

如上所示，teacher 表和 subject 表连接查询，根据 teacher_id 和 extra 字段的值可以看出，subject 表与 teacher 表在连接查询的时候，只要找到一行数据就不再继续查询其他数据。

13.5　MySQL 的索引检索原理

前面已经介绍了索引的创建、索引的指导原则，以及通过分析执行计划来查询是否使用了索引、是否可以优化。接下来将介绍索引查询的底层原理（下面所有的示例都是以 InnoDB 存储引擎下的索引来进行分析的）。

13.5.1　主键索引

主键索引又叫聚簇索引，它使用 B-Tree 构建，叶子节点存储的是数据表的某一行数据。当表没有创建主键索引时，InnoDB 存储引擎会自动创建一个 ROWID（占用 6 字节）字段用于构建聚簇索引。具体规则如下：

（1）在数据表上定义主键，InnoDB 默认会将主键索引用作聚簇索引。

（2）如果数据表没有定义主键，那么 InnoDB 存储引擎会选择第一个不为 null 的唯一索引列作为聚簇索引。

（3）如果以上两个都没有，那么 InnoDB 存储引擎会使用一个 6 字节长整型的隐式字段（ROWID 字段）构建聚簇索引。该 ROWID 字段会在插入新行时自动递增。

下面通过示例来分析主键索引检索数据的逻辑。

例 13.20　主键索引检索数据的逻辑分析。

```
# 创建 student 表
 create table `student`  (
  `id` int(11) not null auto_increment,
  `name` varchar(100) character set utf8 collate utf8_general_ci not null,
  `age` int(11) null default null,
  primary key (`id`) using btree,
  index `index_age`(`age`) using btree
) engine = innodb auto_increment = 66 character set = utf8 collate = utf8_general_ci row_format = compact;
```

通过上面的语句可以看出，id 是主键，也会被默认为主键索引（聚簇索引），而 age 字段被创建为普通索引，这两个字段的索引都是以 B-Tree 的结构存储的。所以目前 student 表的索引树有两棵，一棵是 id 的主键索引树，另一棵是 age 的普通索引树。

id 的主键索引树如图 13-2 所示。

图 13-2　id 的主键索引树

其中，p1、p2 等表示的是存储子节点的地址信息（指针）；45、101 等表示的是主键，也就是 id 字段的值；data 表示主键对应行的数据（当前行的所有字段值）。

age 的普通索引树如图 13-3 所示。

图 13-3　age 的普通索引树

其中，p1、p2 等表示的是存储子节点的地址信息（指针）；55、101 等表示的是索引值，也就是 age 字段的值。

1．主键等值查询

根据上述 id 的主键索引树，我们将解析主键等值查询所经历的过程。

例 13.21　查询 id=39 的行数据，示例 SQL 语句如下：

```
select * from student where id = 39;
```

因为 id 是主键，所以需要从 id 的主键索引树中查询当前行数。

查询过程如下：

- 第一次查询：从根节点检索，将数据块 1 加载到内存中，比较 39<45，遍历左边。
- 第二次查询：将左边的数据块 2 加载到内存中，比较 9<38<39，遍历右边。
- 第三次查询：将右边的数据块 6 加载到内存中，比较 38<39、39=39。查询完毕，将数据返回客户端。

查询流程图如图 13-4 所示。

图 13-4　主键等值的查询流程

2. 主键范围查询

接下来将了解主键范围查询的过程。

例 13.22　查询 id 在 39 和 45 之间的值的行数据,示例 SQL 语句如下:

```
select * from student where id between 39 and 45;
```

因为 id 是主键,所以索引会从主键索引树去检索数据。

查询的过程如下:

- 第一次查询:从根节点检索,将数据块 1 加载到内存中,比较 39<45,遍历左边。
- 第二次查询:将左边的数据块 2 加载到内存中,比较 9<38<39,遍历右边。
- 第三次查询:将右边的数据块 6 加载到内存中,比较 38<39,39=39,遍历右边。
- 第四次查询:将右边的数据块 7 加载到内存中,比较 39<45,45=45。因为叶子节点有双向指针,范围查询可以直接利用双向有序链表,所以查询到这就查询完毕,将数据返回客户端。

查询流程如图 13-5 所示。

图 13-5　主键范围的查询流程

13.5.2　普通索引

本小节将介绍普通索引查询的过程和原理。在 InnoDB 存储引擎中，普通索引不存储行数据，只存储数据的主键值（最底层节点存储主键 id 的值）。比如表 student 中的 age，它的索引结构如图 13-6 所示。

图 13-6　age 索引示例图

如果我们执行语句"select * from student where age = 49;",那么它的查询流程是怎样的呢?

首先我们需要根据 age 索引树查询当前 age=49 所对应的主键 id 的值,然后通过主键 id 的值去主键索引树扫描并查询当前主键 id 具体存放数据的节点,最后返回数据给客户端。

根据 age 查询总共进行了 6 次查询操作,具体查询逻辑如下:

(1)根据 age 索引查询主键 id:

- 第一次查询:从根节点检索,将数据块 1 加载到内存中,比较 49<55,遍历左边。
- 第二次查询:将左边的数据块 2 加载到内存中,比较 29<48<49,遍历右边。
- 第三次查询:将右边的数据块 6 加载到内存中,比较 48<49、49=49,得到主键 39。

(2)根据主键 id 的值查询当前行数据:

- 第四次查询:从根节点检索,将根节点加载到内存中,比较 39 < 45,遍历左边。
- 第五次查询:将左边的数据块 2 加载到内存中,比较 9<38<39,遍历右边。
- 第六次查询:将右边的数据块 6 加载到内存中,比较 38<39、39=39。查询完毕,将数据返回客户端。

查询过程如图 13-7 和图 13-8 所示。

图 13-7　根据 age 索引查询主键 id

图 13-8 根据主键 id 的值查询当前行数据

13.5.3 复合索引

在查询时,如果经常通过几个字段去筛选,则可以把这几个字段结合在一起创建一个复合索引。比如创建学生表的 age 和 name 字段为一个复合索引,索引的结构如图 13-9 所示。

图 13-9 age 和 name 的复合索引结构图

由图 13-9 可以发现,复合索引和普通索引大部分一致,最底层叶子节点存储主键 id 的值。

主要区别是，单个字段的普通索引节点中存储的是一个字段的值；而在复合索引中，节点把字段的值通过逗号隔开存储。执行"select * from student where name = 'libai' and age = 49;"的时候，因为创建的复合索引是 age 和 name，所以索引优化器会调整筛选的顺序，类似把 SQL 优化成"select * from student where age = 49 and name = 'libai'"，然后到复合索引树中去查找当前筛选条件对应的主键 id，最后查询主键索引树，查询到当前行所有的数据。

根据 age 和 name 筛选，总共进行了 6 次查询操作，具体查询逻辑如下：

（1）检索 age 和 name 复合索引树：

- 第一次查询：从根节点检索，将数据块 1 加载到内存中，比较 49<55，遍历左边。
- 第二次查询：将左边的数据块 2 加载到内存中，比较 29<48<49，遍历右边。
- 第三次查询：将右边的数据块 6 加载到内存中，比较 48<49、49=49，得到主键为 39。

（2）检索主键 id 索引树：

- 第四次查询：从根节点检索，将根节点加载到内存中，比较 39<45，遍历左边。
- 第五次查询：将左边的数据块 2 加载到内存中，比较 9<38<39，遍历右边。
- 第六次查询：将右边的数据块 6 加载到内存中，比较 38<39、39=39。查询完毕，将数据返回客户端。

查询过程如图 13-10 和图 13-11 所示。

图 13-10　检索 age 和 name 复合索引树

如图 13-10 所示，经过 3 次查询，找到了当前 age 和 name 值对应的主键 id 值为 39。然后根据主键 39 查询主键索引树，查询当前行的所有数据，如图 13-11 所示。

图 13-11　检索主键 id 索引树

13.6　索引的使用技巧与 SQL 优化实战

本节介绍提升索引的查询效率的技巧及 SQL 优化实战。

1. 避免回表查询

查询数据时，因为每个索引是一棵树，而普通索引叶子节点存储的是主键 id，需要根据主键 id 才能查询到当前行的数据，所以使用普通索引查询时，首先要查询普通索引树来找到主键 id，然后带着主键 id 查询主键索引树，最后在主键索引树的叶子节点中获得数据。我们把这种查询方式叫作回表。如前所述，回表查询需要扫描两棵树进行查询，在性能上明显不如只扫描一棵树的查询。

例 13.23　优化语句避免回表。

```
select * from student where name='wangwu';
```

如上语句所示，假如 name 是普通索引，则先查询普通索引，然后查询主键 id 索引树，这就是回表。在业务允许的情况下，可以通过索引覆盖（查询的所有字段都是索引字段）的方式查询，语句优化如下：

```
select id,name from student where name='wangwu';
```

因为索引叶子节点存储了主键 id，所以通过 name 的索引树即可获取 id 和 name，无须回表。

2. 尽量使用联合索引

在建立索引时，尽量在多个单列索引上判断是否可以使用联合索引。联合索引的使用不仅可以节省空间，还可以更容易地使用索引覆盖。联合索引的创建原则是：在创建联合索引时，应该把频繁使用的列、区分度高的列放在前面，频繁使用代表索引利用率高，区分度高代表筛选粒度大。另外，也可以将经常需要作为查询返回的字段增加到联合索引中。如果在联合索引上增加一个索引字段而能使查询语句返回的字段都是索引字段，那么这种情况下应该使用联合索引。

3. 索引条件下推优化

索引条件下推（Index Condition Pushdown，ICP）是针对 MySQL 使用索引从表中检索行数据的优化。在没有 ICP 的情况下，存储引擎遍历索引以定位基表中的行并将它们返回给 MySQL 服务器，该服务器评估 where 行的条件。启用 ICP，如果 where 可以仅使用索引中的列来评估部分条件，则 MySQL 服务器会把该部分 where 条件推送（下降）到存储引擎。接下来，存储引擎使用索引项评估推送的索引条件，并且仅当满足该条件时才从表中读取行数据。ICP 可以减少存储引擎访问基表的次数和 MySQL 服务器访问存储引擎的次数。启用 ICP 优化性能的示例如下：

```
select * from emp where age=18 and name='caly';
```

在以上语句中，如果不开启 ICP，首先会根据 age=18 来查找记录，检索的结果将指向聚簇索引，最后根据 name='caly'进行过滤，把最终结果返回给用户。开启 ICP 后，在根据 age=18 查找记录的同时会根据 name='caly'进行过滤，然后把检索的结果指向聚簇索引，最后返回给用户。ICP 减少了存储引擎访问表的次数，从而提升了性能。

ICP 开启方式如下：

```
# 关闭索引条件下推（默认开启）
set optimizer_switch = 'index_condition_pushdown=off';
# 开启索引条件下推
set optimizer_switch = 'index_condition_pushdown=on';
```

4. 避免全表扫描

当 MySQL 使用全表扫描查询时，通过 explain 进行分析，如果 type=all，则表示是全表扫描。如果此表有索引并且索引字段在筛选条件中，则可以强制通过索引的方式执行检索，不过这种方式最好经过测试。语法如下：

```
select * from t1, t2 force index (index_for_column)
  where t1.col_name=t2.col_name;
```

5. 负向查询不能使用索引

对于负向查询（!=，not in），因为不能命中索引，从而会进行全表扫描，效率最低，所以在业务中尽量避免进行负向查询。

例 13.24　修改负向查询。

```
select name from emp where id not in (1,3,5);
```

应该修改为：

```
select name from emp where id in (2,4,6);
```

6. 前导模糊查询不能使用索引

对于前导模糊查询，MySQL 在查询的时候不会使用索引进行查询，所以其查询效率较低。在业务允许的情况下，尽量避免使用前导模糊查询。

例 13.25　修改前导模糊查询。

```
select name from emp where name like '%zhangsan'
```

应该修改为：

```
select name from emp where name like 'zhangsan%'
```

7. 在字段上计算不能命中索引

在查询过程中，如果对索引字段进行函数计算，则在查询过程中不会通过索引进行查询，查询性能较低，所以在业务允许的情况下，尽量避免对索引字段进行函数计算。

例 13.26　修改索引字段的函数计算。

```
select name from emp where from_unixtime(create_time) < curdate();
```

应该修改为：

```
select name from emp where create_time < from_unixtime(curdate());
```

8. 查询条件类型不一致不能命中索引

在查询过程中，对字段进行筛选查询，如果字段自身的类型和筛选内容的类型不匹配，则会导致查询不会进行索引，查询性能较低，所以在业务中需要避免出现此类情况。

例 13.27　查询条件中的类型应该保持一致（emp 中 id 字段的类型是 int 类型）。

```
select name from emp where id='12';
```

应该修改为：

```
select name from emp where id=12;
```

9. 明确知道只返回一条记录时可以使用 limit 进行限制

在查询过程中，如果明确知道查询的结果只有一条记录，则可以使用 limit 进行限制，这会在很大程度上提高查询的效率。

例 13.28 明确知道只返回一条记录时可以使用 limit 进行限制。

```
select name from emp where username='zhangsan' limit 1
```

10. limit 分页优化

在了解 limit 分页优化之前，我们先看如下的 SQL 语句：

```
select * from test1 order by id limit 99999,10;
```

该 SQL 虽然用到了 id 索引，但是从第一行起要定位到第 99999 行，再向后扫描 10 行，等于在做全表扫描，其性能非常低下。优化后的 SQL 语句如下：

```
 select * from test1 order by id where id>=100000 order by id limit 10;
```

该 SQL 语句利用 id 索引直接定位到第 100000 行，再向后扫描 10 行，相当于在一个范围内扫描，其性能提升不少。

11. 使用表内连接进行 limit 分页优化

在了解使用表内连接进行 limit 分页优化之前，我们先看如下的 SQL 语句：

```
Select id,title,createdate from test1 order by createdate asc limit 100000,10;
```

优化后的 SQL 语句如下：

```
 Select a.id, a.title,a.createdate from test1 a join (select id from test1 order by createdate asc limit 100000,1) b on a.id>=b.id limit 10;
```

优化的思路是：先取出 99999 行后面的一条记录的 id，再用表内连接的方法取出后面 10 条。

12. union all 优化（基于 MySQL 8.0+）

在 MySQL 5.7 中，union all 不再创建一张临时表，这样在执行大的联合查询时会减少 I/O 开销，从而提升查询速度，但对 union 语句和在最外层使用的 order by 语句无效。如下 SQL 语句将不会产生临时表。

```
(select id from t1 order by id) union all (select id from t2 where k=1 order by id);
```

但是，如下 SQL 语句会产生临时表：

```
(select id from t1 order by id) union (select id from t2 where k=1 order by id);
(select id from t1 order by id) union all (select id from t2 where k=1 order by id) order by id desc;
```

13. or 优化（基于 MySQL 8.0+）

首先查看如下的 SQL 语句：

```
select * from city where category='A' or category='B'
```

如上 SQL 语句虽然在 category 字段上加了索引，但是索引使用的是范围类型，如果将索引使用类型优化成 ref 类型，则查询效率将有所提高。下面使用 union all 对使用 or 的语句进行优化（优化之后使用的是 ref 类型）。

```
(select * from city where category='A') union all (Select * from city where category='B' )
```

14. count(*)优化

首先查看如下 SQL 语句：

```
select count(*) from userinfo;
```

优化后的 SQL 语句如下：

```
Select count(*) from up_user where sid>=0;
```

这样优化后，可以不需要扫描全表而通过扫描索引就能查询出结果。当表数据量很大时，效率会明显提高。

15. on duplicate key update 优化

MySQL 有一种高效的主键冲突判断功能，即冲突时执行 update 逻辑，不冲突时执行 insert 逻辑，语法如下（这种操作必须基于主键或者唯一索引的操作）：

```
INSERT INTO table_name (col1, col2, col3, ...) VALUES (val1, val2, val3, ...)
ON DUPLICATE KEY UPDATE col1 = val1_updated, col2 = val2_updated, col3 = val3_updated, ...;
```

其中，table_name 为目标表名，(col1, col2, col3, ...)则是要插入的列名称，(val1, val2, val3, ...)是对应的值。如果这些列中的某个列定义了主键或唯一索引，并且插入的数据与表中已有的数据具有相同的值，则将执行 UPDATE 语句更新现有记录的列。在 UPDATE 子句中，可以指定要更新的列及其新值。如果没有冲突，则将像通常情况一样执行 INSERT 操作。需要注意的是，在执行 UPDATE 时，MySQL 将使用新插入数据中的值更新与重复键匹配的任何现有行。如果我们需要在不同列上具有唯一性索引，那么必须使用组合索引才能避免冲突，或者使用 replace into 语法在发生冲突时直接替换原有记录，但它将删除并替换现有记录，而不仅仅是更新记录的特定列。

例 13.29 on duplicate key update 优化的示例 SQL 语句及其执行结果如下：

```
mysql> create table t (c1 char(5), c2 varchar(10), c3 text, c4 blob);
Query OK, 0 rows affected (0.00 sec)
mysql> delimiter ;
```

```
mysql> alter table t add column c5 blob default '30' on duplicate key update
c2=values(c2),c3=values(c3),c4=values(c4);
Query OK, 1 row affected (0.00 sec)
mysql> select * from t order by c1;
+-----+-----+------+------+---------+
| c1  | c2  | c3   | c4   | c5      |
+-----+-----+------+------+---------+
|     |     | NULL | NULL |         |
|     |     | NULL | NULL |         |
|     |     | NULL | NULL |         |
|     |     | NULL | NULL |         |
|     |     | NULL | NULL |         |
|     |     | NULL | NULL |         |
+-----+-----+------+------+---------+
5 rows in set (0.00 sec)
```

以上 SQL 代码创建了一张名为 t 的表，其中包含四个字符列和一个二进制列。在第一个查询中，我们打印了表 t 中的记录，并按照字符列 c1 的顺序排序。由于字符列 c1 具有默认值为空字符串，因此它出现在输出结果的第一位。在第二个查询中，我们向表 t 中添加了一个二进制列 c5，并将其默认值设置为字符串"30"，同时指定了 on duplicate key update 子句，以便在主键或唯一索引发生冲突时执行更新逻辑。最后，我们打印了表 t 中的记录，并按照字符列 c1 进行排序。

13.7 本 章 练 习

1. 说说你对索引的认识。
2. 写出创建主键索引、普通索引、复合索引的语法，并说明它们各自的使用特点。
3. 举例说明联合索引的使用场景。
4. 假设你是一名程序员，在程序应用过程中有一个业务突然出现请求效率变慢的情况，你将如何解决这个问题来提升此业务的请求效率。

第 14 章

InnoDB 存储引擎事务模型和锁的使用

本章内容：

- InnoDB 存储引擎的实践
- InnoDB 和 ACID 模型
- InnoDB MVCC 实现原理
- 各种锁原理解析
- 事务隔离级别

　　本章将讲解 MySQL 中 InnoDB 存储引擎的核心内容。如果想要在工作中提升自己的业务能力，那么 InnoDB 的内存结构和优化细节、数据库中各种锁的应用及其原理都是必须掌握的知识。本章会列举大量示例供读者理解和学习。

　　InnoDB 是一种兼顾高可靠性和高性能的通用存储引擎。在 MySQL 8.0 中，InnoDB 是默认的 MySQL 存储引擎，除非用户配置了不同的存储引擎作为默认存储引擎。InnoDB 存储引擎主要有下面几个优势：

　　（1）DML操作遵循ACID（事务）模型，事务具有提交、回滚和崩溃恢复功能，以保护用户数据。

　　（2）行级锁定和Oracle风格一致的读取方式提高了多用户并发性和性能。

　　（3）InnoDB表根据主键优化查询。每张InnoDB表都有一个称为聚簇索引的主键索引，该索引数据用于最小化I/O查询。

　　（4）为了维护数据的完整性，InnoDB支持外键约束。使用外键检查插入、更新和删除操作，以确保这些操作不会导致相关表数据之间的不一致。表14-1是对InnoDB存储引擎特征的说明。

表 14-1 InnoDB 存储引擎特征

特　　征	是 否 支 持
B-Tree 索引	是
备份/时间点恢复	是
集群数据库支持	否
聚簇索引	是
压缩数据	是
数据缓存	是
加密数据	是（加密函数在服务器中的实现。在 MySQL 5.7 及更高版本中，支持静态数据加密）
外键支持	是
全文检索索引	是（MySQL 5.6 及更高版本支持全文索引）
地理空间数据类型支持	是
地理空间索引支持	是（MySQL 5.7 及更高版本支持地理空间索引）
哈希索引	否（InnoDB 在内部利用哈希索引来实现其自适应哈希索引功能）
索引缓存	是
MVCC	是
复制支持	是
存储限制	是（最大支持 64TB 存储）

InnoDB 存储引擎有如下优点：

（1）如果服务器由于硬件或软件问题而意外退出，那么无论当时数据库中发生了什么，我们在重新启动数据库后都不需要执行任何特殊操作。InnoDB具备崩溃恢复功能，它会自动完成在崩溃之前提交的所有更改，并撤销正在进行但未提交的更改，使得在重新启动后可以从上次中断的地方继续进行后续操作。

（2）InnoDB存储引擎维护自己的缓冲池，把经常使用的数据存放在内存中，用于主内存缓存表和索引数据的查询。在专用数据库的服务器上，多达80%的物理内存通常分配给缓冲池。

（3）InnoDB存储引擎的表支持外键。

（4）InnoDB存储引擎的校验机制可以检测到磁盘或内存中的数据损坏。

（5）当为数据库的表设置主键列时，系统会自动优化主键列，在where子句、order by子句、group by子句和连接操作中提高主键的使用效率。

（6）InnoDB存储引擎不仅允许对同一张表进行并发读写访问，它还缓存更改的数据以减少对磁盘的I/O操作。

（7）当从表中重复访问相同的行数据时，自适应哈希索引会接管这些数据，以提高查询效率。

（8）可以压缩表和索引数据，以及加密表数据。

（9）支持在线 DDL 操作。

（10）支持通过查询 information_schema 表来监控存储引擎的内部工作。

（11）支持通过查询 performance_schema 表来监控存储引擎性能的详细信息。

（12）将 InnoDB 表与来自其他 MySQL 存储引擎的表混合使用。

（13）InnoDB 存储引擎提高了处理大量数据时的 CPU 效率。

（14）InnoDB存储引擎即使在文件大小限制为2GB的操作系统上也可以操作大数据量的表。

14.1　InnoDB 存储引擎实践

下面将介绍使用 InnoDB 存储引擎时的推荐做法。

（1）在查询时最好根据主键查询，如果没有主键，可以指定一个自增值作为主键。

（2）在使用连接查询时，最好在连接列上定义外键，而且表中的连接列使用相同的数据类型声明，这样可以提高查询性能。外键还会将删除和更新操作传播到所有受影响的表中，并在父表中不存在相应id时阻止在子表中插入数据，从而保证数据的完整性。

（3）关闭自动提交。每秒提交数百次会限制服务性能，可以执行"set autocommit=0;"进行关闭。

（4）通过用start transaction和commit语句将相关的DML操作集分组到事务中。

（5）不要使用lock tables语句。InnoDB存储引擎允许多个会话同时对同一张表进行读取和写入，而不会影响可靠性和高性能。如果需要获得对一行数据的独占写访问权限，可以使用select … for update语法锁定待更新的行。

（6）启用innodb_file_per_table变量或使用通用表空间将表内的数据和索引放入单独的文件中。innodb_file_per_table 默认情况下为启用状态。

（7）可以在InnoDB存储引擎不牺牲读写能力的情况下压缩表。

（8）为了防止启用用户不想使用的存储引擎来创建表，可以使用sql_mode=no_engine_substitution选项来运行服务器。

如果想要知道当前服务器的默认存储引擎，则可以使用下面的 SQL 语句来查看：

```
show engines;
```

查询结果如图 14-1 所示。

```
mysql> SHOW ENGINES;
+--------------------+---------+----------------------------------------------------------------+--------------+------+------------+
| Engine             | Support | Comment                                                        | Transactions | XA   | Savepoints |
+--------------------+---------+----------------------------------------------------------------+--------------+------+------------+
| InnoDB             | DEFAULT | Supports transactions, row-level locking, and foreign keys     | YES          | YES  | YES        |
| MRG_MYISAM         | YES     | Collection of identical MyISAM tables                          | NO           | NO   | NO         |
| MEMORY             | YES     | Hash based, stored in memory, useful for temporary tables      | NO           | NO   | NO         |
| BLACKHOLE          | YES     | /dev/null storage engine (anything you write to it disappears) | NO           | NO   | NO         |
| MyISAM             | YES     | MyISAM storage engine                                          | NO           | NO   | NO         |
| CSV                | YES     | CSV storage engine                                             | NO           | NO   | NO         |
| ARCHIVE            | YES     | Archive storage engine                                         | NO           | NO   | NO         |
| PERFORMANCE_SCHEMA | YES     | Performance Schema                                             | NO           | NO   | NO         |
| FEDERATED          | NO      | Federated MySQL storage engine                                 | NULL         | NULL | NULL       |
+--------------------+---------+----------------------------------------------------------------+--------------+------+------------+
9 rows in set (0.00 sec)
```

图 14-1　查看默认的存储引擎

14.2　InnoDB 和 ACID 模型

ACID（事务）模型是一组数据库设计原则，是业务数据和关键任务应用的可靠性保证。在 MySQL 中，InnoDB 存储引擎是严格遵守 ACID 模型的存储引擎，因此数据不会损坏，执行数据的结果不会因软件崩溃或硬件故障等异常情况而失真。当数据本身符合 ACID 的特性时，应用程序不需要重新发明一致性检查和崩溃恢复机制的轮子。如果有额外的软件保护措施、超可靠的硬件或可以容忍少量数据丢失的应用程序，则可以调整 MySQL 设置以获得更高的性能或吞吐量。下面我们将介绍 InnoDB 存储引擎的 ACID 模型。

MySQL 事务主要用于处理操作量大、复杂度高的数据。比如，在人员管理系统中删除一个人员，既要删除人员的基本资料，又要删除和该人员相关的信息，如信箱、文章等，如果其中有一项没有删除成功，则其他内容也不能被删除。这样，这些数据库操作语句就构成了一个事务。

一般来说，事务必须满足 4 个条件（ACID）：原子性（Atomicity，或称为不可分割性）、一致性（Consistency）、隔离性（Isolation，又称为独立性）、持久性（Durability）。

- 原子性：一个事务中的所有操作可以全部完成或全部失败，但不会结束在中间某个环节。事务在执行过程中发生错误，会被回滚（Rollback）到事务开始前的状态，就像这个事务从来没有执行过一样。如图 14-2 所示，用户 A 给用户 B 转账 1000，A 账户减去 1000，B 账户加上 1000，这个操作是一个原子操作，是不可分割的。如果同时成功，则数据库中 A 账户减去 1000，B 账户加上 1000；如果同时失败，则操作回滚，数据库中的数据不变。

图 14-2　转账业务

- 一致性：在事务开始之前和事务结束以后，数据库的完整性不会被破坏。这表示写入的数据必须完全符合所有的预设规则，包含数据的精确度、串联性，以及后续数据库可以自发性地完成预定的工作。
- 隔离性：数据库允许多个并发事务同时对其数据进行读写和修改，隔离性可以防止多个事务并发执行时由于交叉执行而导致数据不一致。事务隔离分为不同级别，包括未提交读（read uncommitted）、提交读（read committed）、可重复读（repeatable read）和串行化（serializable）。
- 持久性：事务处理结束后，对数据的修改就是永久的，即使系统故障也不会丢失。

> **提 示**
>
> 在 MySQL 命令行的默认设置下，事务都是自动提交的，即执行 SQL 语句后就会马上执行 commit 操作。因此，要显式地开启一个事务需要使用命令 begin 或 start transaction，或者执行命令 set autocommit=0 来禁用当前会话的自动提交。

14.2.1　MySQL 事务处理的两种方法

1. 用 begin、rollback、commit 实现

- begin：开始一个事务。
- rollback：事务回滚。
- commit：事务确认。

2. 直接用 set 来改变 MySQL 的自动提交模式

- set autocommit=0：禁止自动提交。
- set autocommit=1：开启自动提交。

例 14.1　MySQL 事务处理操作的示例 SQL 语句及其执行结果如下：

```
mysql> use clay;
Database changed
# 创建数据表
mysql> create table clay_transaction_test( id int(5)) engine=innodb;
Query OK, 0 rows affected (0.04 sec)

mysql> select * from clay_transaction_test;
Empty set (0.01 sec)
# 开始事务
mysql> begin;
Query OK, 0 rows affected (0.00 sec)

mysql> insert into clay_transaction_test value(1);
Query OK, 1 rows affected (0.01 sec)

mysql> insert into clay_transaction_test value(2);
Query OK, 1 rows affected (0.00 sec)
# 提交事务
```

```
mysql> commit;
Query OK, 0 rows affected (0.01 sec)
mysql> select * from clay_transaction_test;
+------+
| id   |
+------+
| 1    |
| 2    |
+------+
2 rows in set (0.01 sec)
# 开始事务
mysql> begin;
Query OK, 0 rows affected (0.00 sec)

mysql> insert into clay_transaction_test values(3);
Query OK, 1 rows affected (0.00 sec)
# 回滚
mysql> rollback;  Query OK, 0 rows affected (0.00 sec)
# 因为回滚，所以数据没有插入
mysql> select * from clay_transaction_test;
+------+
| id   |
+------+
| 1    |
| 2    |
+------+
2 rows in set (0.01 sec)
```

14.2.2　InnoDB MVCC

InnoDB 是一个多版本并发控制的存储引擎。它保留与已更改数据行旧版本有关的信息以支持事务功能，例如并发和回滚，该信息以回滚段的数据结构存储在撤销表的空间中。InnoDB 使用回滚段中的信息来执行事务回滚所需的撤销操作。这些信息还用来构建数据行的早期版本以实现一致读取。在 InnoDB 内部，InnoDB 为存储在数据库中的每一行添加 3 个字段：

（1）一个 6 字节的 DB_TRX_ID 字段，指示插入或更新行的最后一个事务的事务标识符。此外，删除在内部被视为更新，在其中设置数据行中的特殊位以将它标记为已删除。

（2）一个 7 字节的 DB_ROLL_PTR 字段，被称为回滚指针。回滚指针指向写入回滚段的撤销日志记录。如果该行被更新，则撤销日志记录该行更新前重建内容所需的所有信息。

（3）一个 6 字节的 DB_ROW_ID 字段，存储一个行的 id 值，该值随着新行的插入而递增。如果 InnoDB 自动生成聚簇索引，则该索引包含行的 id 值；否则该 DB_ROW_ID 列不会出现在任何索引中。

回滚段中的撤销日志分为插入撤销日志和更新撤销日志。

- 插入撤销日志仅在事务回滚时进行，并且可以在事务提交后立即丢弃。
- 更新撤销日志也用于一致性读取，但是只有在没有事务存在且为其 InnoDB 分配快照的情况下才能丢弃它们，在一致性读取中可能需要更新撤销日志中的信息来构建较早版本的数据库数据。

回滚段中的撤销日志记录的数据物理大小通常小于相应的插入或更新的行数据。用户可以依据此信息来计算回滚段中所需的空间。

在 InnoDB MVCC（Multi-Version Concurrency Control，多版本并发控制）方案中，当用户使用 SQL 语句删除某行时，不会立即从数据库中物理地删除该行。InnoDB 只有在删除撤销日志记录（delete 操作写入的撤销日志记录）时，才会物理地删除相应的行及其索引记录。这种删除操作称为清除，执行速度非常快，通常与执行删除操作的 SQL 语句花费的时间相同。

14.3 锁机制

数据是一种供许多用户共享访问的资源，如何保证数据库并发访问的一致性、有效性是所有数据库必须解决的一个问题，锁的冲突也是影响数据库并发访问性能的一个重要因素。从这个角度来说，锁对于数据库而言就显得尤为重要。MySQL 中的锁可以按照不同的维度进行分类，以下是常见的几种锁：

（1）按照粒度分类：

- 表级锁：锁住整张表，其他会话不能对特定行进行修改。
- 行级锁：在某些情况下，MySQL 会将锁的粒度进行缩小，锁住某几行。行级锁比表级锁更细粒度，其他会话仍然可以修改表中其他行。

（2）按照类型分类：

- 共享锁（S 锁或者读锁）：多个会话可以同时对同一份数据对象进行共享锁定，读取数据对象的操作都是可行的，但写入数据对象的操作则会被阻塞。
- 排他锁（X 锁或者写锁）：只有一个会话可以对一个数据对象进行排他锁定，其他会话不能进行读取或者写入数据对象的操作，只能等待排他锁被释放。
- 意向锁：用于在同一个事务的行级锁和表级锁之间进行转换和协调，降低锁定的粒度并提高并发性能。
- 乐观锁：假定并发操作之间没有冲突，不会阻塞数据读写操作，只在提交数据时进行数据版本的校验和冲突检测，如果存在冲突则回滚操作。
- 悲观锁：默认并发操作之间存在冲突，通过锁机制保护数据对象的完整性，会降低并发性能。
- 间隙锁：锁住一个数据对象前后的间隔区域以避免幻读问题。
- 记录锁：锁住某个或某些行，但是不包含间隙。
- 临建锁：锁住临时表，例如 MySQL 的 temporary table。
- 死锁：两个或多个事务互相需要对方占有的资源而陷入的循环等待状态。

需要注意的是，这些分类是有重叠的，例如一把排他锁也可以是行级锁。分类方式的选择取决于具体的场景需求和实现细节，避免锁对系统性能造成负面影响。

14.3.1 表级锁

表级锁是 MySQL 锁中粒度最大的一种锁，表示当前的操作对整张数据表加锁，资源开销比行级锁少，不会出现死锁的情况，但是发生锁冲突的概率很大。该锁的锁定机制最大的特点是实现逻辑非常简单，带来的系统负面影响最小，所以获取锁和释放锁的速度很快。由于表级锁一次会将整张数据表锁定，因此可以很好地避免困扰我们的死锁问题。InnoDB 支持表级锁，但是默认使用的是行级锁，而且只有在查询或者其他 SQL 语句通过索引操作时才会使用行级锁。

在 MySQL 中，可以通过以下两种方式实现表级锁：

- 共享锁：通过给表加上共享锁来控制并发读操作。例如，可以使用 select 语句来获取表中的数据，并使用 select...for update 语句来获取共享锁。
- 排他锁（exclusive lock）：通过给表加上排他锁来控制并发写操作。例如，可以使用 select...lock in share mode 语句来获取排他锁。需要注意的是，在使用排他锁时，需要指定 lock in share mode 语句，以确保其他事务不能同时获取该行数据的排他锁。

14.3.2 行级锁

行级锁是 MySQL 锁中粒度最小的一种锁，因为锁的粒度最小，所以发生资源争抢的概率也最小，并发性能最大，但是可能会造成死锁，每次加锁和释放锁的开销也会变大。目前主要是 InnoDB 使用行级锁。

行级锁可以通过以下两种方式实现：

- 基于悲观锁实现：在事务访问数据时，将所访问的数据行加上排他锁，让其他事务无法修改或读取该行数据。
- 基于乐观锁实现：在事务读取数据时，不加锁并记录当前数据的版本号或时间戳，当需要更新数据时，检查数据的版本号或时间戳是否与事务开始时一致，一致则说明没有其他事务修改，将变更更新到数据库中，不一致则说明该数据被其他事务修改，事务回滚并重试操作。

在使用行级锁时需要注意以下几点：

（1）行级锁降低了并发性，但提高了数据的一致性，使用时需要充分考虑业务对并发性和一致性的需求。

（2）行级锁可以在修改、删除、插入时使用，但对于频繁读取的数据，也可以考虑使用乐观锁来实现，避免锁的开销。

（3）行级锁可能产生死锁问题，在实现时需要注意避免。

（4）MySQL 默认实现的行级锁是基于 InnoDB 存储引擎，其他存储引擎可能支持不同的锁定机制。

总之，行级锁在需要精细控制事务并发的业务场景下是非常有用的。

14.3.3 共享锁

共享锁的具体逻辑为：若事务 A 对数据对象 o 加上 S 锁，则事务 A 可以读数据对象 o，但不能修改，其他事务只能再对数据对象 o 加 S 锁，而不能加 X 锁，直到事务 A 释放数据对象 o 上的 S 锁。这样保证了其他事务可以读数据对象 o，但在事务 A 释放数据对象 o 上的 S 锁之前，不能对数据对象 o 进行任何修改。

共享锁的语法如下：

```
# 加共享锁
lock table tablename read;
# 释放锁
unlock table;
```

例 14.2 给数据表创建共享锁的示例 SQL 语句及其执行结果如下：

```
# 在session1 上执行，给数据表创建读锁（创建一个连接就是一个会话）
mysql> lock table userinfo read;
Query OK, 0 rows affected (0.01 sec)
# 可以读取当前表的数据
mysql> select * from userinfo ;
+----+----------+------+
| id | name     | age  |
+----+----------+------+
|  1 | zhangsan |   27 |
|  2 | lisi     |   27 |
|  3 | dd       | NULL |
|  4 | dd       |    2 |
+----+----------+------+
4 rows in set (0.00 sec)
# 在session1 上执行，因为给 userinfo 表加了读锁，所以不能读取其他表数据
mysql> select * from student_score;
ERROR 1100 (HY000): Table 'student_score' was not locked with LOCK TABLES
# 在session1 上执行， 因为给表加的读锁，所以当前会话不能对表执行其他操作
mysql> update userinfo set age=age+1;
ERROR 1099 (HY000): Table 'userinfo' was locked with a READ lock and can't be updated

# 在session2 上执行，在session1 没有释放锁之前，session2 可以读取 userinfo 表
mysql> select * from userinfo;
+----+----------+------+
| id | name     | age  |
+----+----------+------+
|  1 | zhangsan |   27 |
|  2 | lisi     |   27 |
|  3 | dd       | null |
|  4 | dd       |    2 |
+----+----------+------+
4 rows in set (0.00 sec)
# 在session2 上执行，在session1 没有释放锁之前，session2 修改或删除表 userinfo 的任何数据都会被阻塞
mysql> update  userinfo set age=1 where name='dd';
```

```
# 一直等待，直到 seesion1 释放锁
# 在 session1 上执行，释放锁，session1 在执行下面释放锁语句后，session2 才会提交上面的修改语句
mysql> unlock tables;
```

14.3.4 排他锁

排他锁的具体逻辑为：若事务 A 对数据对象 o 加上 X 锁，事务 A 可以读数据对象 o，也可以修改数据对象 o，其他事务则不能再对数据对象 o 加任何锁，直到事务 A 释放数据对象 o 上的锁。这样保证了其他事务在事务 A 释放数据对象 o 上的锁之前不能再读取和修改数据对象 o。

排他锁的语法如下：

```
# 给表加排他锁
Lock table tablename write;
# 给行加排他锁
select ... for update;
# 释放表锁
unlock table;
```

例 14.3　给表加排他锁的示例 SQL 语句及其执行结果如下：

```
# 在 session1 上执行，当前会话给表加写锁
mysql> lock table userinfo write;
Query OK, 0 rows affected (0.00 sec)
# 在 session1 上执行，当前会话可以修改此表数据
mysql> update userinfo set age=9;
Query OK, 4 rows affected (0.01 sec)
Rows matched: 4  Changed: 4  Warnings: 0
# 在 session1 上执行，当前会话可以查询此表
mysql> select * from userinfo;
+----+----------+------+
| id | name     | age  |
+----+----------+------+
|  1 | zhangsan |    9 |
|  2 | lisi     |    9 |
|  3 | dd       |    9 |
|  4 | dd       |    9 |
+----+----------+------+
4 rows in set (0.00 sec)
# 在 session2 上执行，在 session1 没有释放锁之前，session2 对 userinfo 表的任何数据进行操作
（增、删、改、查）都将被阻塞
    mysql> update  userinfo set age=1 where name='dd';
    # 一直等待，直到 seesion1 释放锁
    # 在 session1 上执行，此语句表示释放 userinfo 表的锁
mysql> unlock userinfo ;
    # 当执行上面语句之后，session2 上显示如下执行结果
    2 rows in set (17.10 sec)
```

上面是锁表，接下来我们来看给行加锁的效果。

例 14.4　给行加排他锁的示例 SQL 语句及其执行结果如下：

```
# 修改提交模式，改为手动提交
```

```
mysql> set autocommit=0;
Query OK, 0 rows affected (0.00 sec)
# 在 session1 上执行，当前会话给 id=1 的行加写锁
mysql> select * from userinfo where id=1 for update;
+----+----------+------+
| id | name     | age  |
+----+----------+------+
|  1 | zhangsan |   10 |
+----+----------+------+
1 row in set (0.00 sec)
# 在 session2 上执行，当前会话可以查询当前表的数据
mysql> select * from userinfo;
+----+----------+------+
| id | name     | age  |
+----+----------+------+
|  1 | zhangsan |   11 |
|  2 | lisi     |   12 |
|  3 | dd       |   12 |
|  4 | dd       |   11 |
+----+----------+------+
4 rows in set (0.00 sec)
# 在 session2 上执行，当前会话可以操作其他行数据
mysql> update userinfo set age=12 where id=2;
Query OK, 0 rows affected (0.01 sec)
Rows matched: 1  Changed: 0  Warnings: 0
# 在 session2 上执行，当前会话修改数据涉及 id=1 的行，因此一直被阻塞
mysql> update userinfo set age=12 ;
```

一直被阻塞，直到 sesssion1 执行 commit 语句：

```
# 在 session1 上执行，当前会话提交数据，也就是释放锁
mysql> commit;
Query OK, 0 rows affected (0.00 sec)
```

14.3.5　意向锁

意向共享锁和意向排他锁总称为意向锁。意向锁的出现是为了使 InnoDB 支持多粒度锁，它是表级别的锁。两种锁的含义如下：

- 意向共享锁：事务想要获得表中某些记录的共享锁（读锁），需要在表上先加意向共享锁。
- 意向排他锁：事务想要获得表中某些记录的排他锁（写锁），需要在表上先加意向排他锁。

要给一个表加锁，需要根据意向锁去判断表中有没有数据行被锁定，以确定是否能加锁成功。如果意向锁是行级锁，就得遍历表中所有数据行来判断；如果意向锁是表级锁，则直接判断一次就知道表中是否有数据行被锁定了。因此，将意向锁设置成表级锁的性能比设置成行级锁的性能高得多。

当一个事务需要锁定某个资源时，如果该资源正被一个共享锁锁定，则事务可以再加一个共享锁，不过不能加排他锁；如果该资源正被一个排他锁锁定，则事务只能等待该锁释放资

源之后才能获取锁定资源并添加自己的锁。意向锁的作用是当一个事务在需要获取资源的锁定时，如果该资源已经被排他锁锁定，则数据库会自动给该事务申请一个该表的意向锁，如果事务需要一个共享锁，就申请一个意向共享锁；如果需要的是某行（或者某些行）的排他锁，则申请一个意向排他锁。意向共享锁可以同时并存多个，但是意向排他锁只能存在一个。

操作流程如下：

有了意向锁之后，事务 A 在申请行级别的写锁之前，数据库会自动先给事务 A 申请表的意向排他锁。此时，若事务 B 去申请表的写锁就会失败，因为表上有意向排他锁之后，事务 B 在申请表的写锁时就会被阻塞，这样能够明显提高性能，快速判断出此表是否有被锁的数据。同样也不能获取到此表的表级别的锁。

14.3.6 乐观锁

乐观锁不是数据库自带的，需要我们自己在编写业务代码时去实现。乐观锁是指操作数据库时（更新操作）想法很乐观，认为这次的操作不会导致冲突，在操作数据时不进行任何其他的特殊处理（也就是不加锁），而在进行数据更新后再去判断是否发生了冲突。

操作流程如下：

在表中的数据进行更新操作时，先给数据表加一个版本（version）字段，每操作一次，就将那条记录的版本号加 1，也就是先查询出那条记录，获取 version 字段。如果要对那条记录进行操作（更新），则先判断此刻的 version 值是否与刚刚查询出来的 version 值相等，如果相等，则说明期间没有其他程序对它进行操作，可以执行更新操作，同时将 version 字段的值加 1；如果此刻的 version 值与刚刚获取的 version 值不相等，则说明期间已经有其他程序对它进行操作了，将不进行更新操作。

这样做的好处是避免了长事务中的数据库加锁和解锁的开销，大大提升了大并发量下系统整体的性能表现。

例 14.5 实现乐观锁业务的示例 SQL 语句及其执行结果如下：

```
# 首先获取修改之前的数据和版本号
select data as old_data, version as old_version from …;
# 进行修改，条件是当前数据的版本号和之前的旧版本号一致，表示没有人进行过修改
int updaterow= update set data = new_data, version = new_version where version = old_version
// 表示修改成功，拿到了乐观锁
if (updated row > 0) {
// 乐观锁获取成功，操作完成
} else {
// 乐观锁获取失败，回滚并重试
}
```

14.3.7 悲观锁

悲观锁指的是对数据被外界（包括本系统当前的其他事务，以及来自外部系统的事务处理）修改持保守态度，因此在整个数据处理过程中，数据处于锁定状态。悲观锁的实现往往依靠数据库提供的锁机制（只有数据库提供的锁机制才能真正保证数据访问的排他性，否则即使在本系统中实现了加锁机制，也无法保证外部系统不会修改数据）。

操作流程如下：

在事务 A 中操作数据 1 时，认为一定会有事务 B 来修改数据 1，因此在将数据 1 查询出来后直接加上排他锁（X 锁），防止别的事务来修改数据 1，直到提交后才释放排他锁。

例 14.6 实现悲观锁的示例 SQL 语句及其执行结果如下：

```
# 关闭自动提交
set autocommit=0;
# 开始事务
start transaction;
# 查询当前商品信息，然后通过 for update 锁定数据，防止其他事务修改
select status from goods where goodsid=1 for update;
# 根据商品信息生成订单
insert into orders (id,goodsid) values (null,1);
# 修改商品 status 为 2
update t_goods set status=2;
# 执行完毕，提交事务
commit;
```

> **提　示**
>
> 乐观锁保证了数据处理时的安全性，但是更新有可能会失败，甚至是更新几次都失败。因此，如果写入居多，对吞吐量要求不高，可以使用悲观锁。但是悲观锁加锁会造成开销增加、性能降低，甚至还可能会出现死锁的情况。

14.3.8 间隙锁

间隙锁是 InnoDB 在提交时为了解决幻读问题而引入的锁机制（下面的所有示例没有特意强调都使用可重复读隔离级别）。幻读问题是指在执行新增或者更新操作时，对涉及的数据行进行范围查询（加锁查询）时出现的数据不一致的问题。这时使用不同的行级锁已经没有办法满足要求了，需要对一定范围内的数据进行加锁，间隙锁就用于解决这类问题。在可重复读隔离级别下，数据库是通过行级锁和间隙锁共同组成的 next-key lock 来实现的。加锁规则具有以下特性：

- 加锁的基本单位是 next-key lock，使用前开后闭原则。
- 数据插入过程中访问的对象会增加锁。
- 索引上的等值查询：给唯一索引加锁时，next-key lock 升级为行级锁。
- 索引上的等值查询：向右遍历时，最后一个值不满足查询需求时，next-key lock 退化为间隙锁。
- 唯一索引上的范围查询会访问到不满足条件的第一个值为止。

例 14.7 间隙锁示例。

以表 14-2 所示的数据进行操作时，为了解决幻读问题，更新的时候不只对表中的 5 条数据增加行级锁，还要对中间的取值范围增加间隙锁，间隙锁的取值范围是(–∞,5]、(5,10]、(10,15]、(15,20]、(20,25]、(25,+∞)。

表 14-2 提供的数据

id（主键）	name	age
5	name1	18
10	name2	19
15	name3	20
20	name4	21
25	name5	22

示例 SQL 语句及其执行结果如下：

```
# 在 session1 上执行，操作开启事务
mysql> begin;
Query OK, 0 rows affected (0.00 sec)
# 在 session1 上执行，更新 id=11 的数据，获取行锁
mysql> select * from userinfo where id=11  for update;
Empty set (0.00 sec)

# 在 session2 上执行，操作开启事务
mysql> begin;
Query OK, 0 rows affected (0.00 sec)
# 在 session2 上执行，新增主键为 12 的数据
mysql> insert into userinfo values(12,'hh',18);
...
# 一直处于阻塞状态
# 如果等待时间过长，session1 没有释放锁，则会抛出如下异常
ERROR 1205 (HY000): Lock wait timeout exceeded; try restarting transaction
```

事务 session1 和事务 session2 同时操作时，事务 session1 会对数据库的数据表增加(10, 15]这个区间锁，这时新增 id=12 的数据就会因为区间锁(10, 15]而被锁住无法执行。

需要特别注意的是，如果间隙锁操作不当就会出现死锁。

例 14.8 间隙锁操作不当导致死锁，如表 14-3 所示。

表14-3 死锁示例

事务 A	事务 B
begin; select * from t where id = 9 for update;	
	begin; select * from t where id = 6 for update; insert into user value(7,7,7) 一直阻塞
insert into user value(7,7,7) 一直阻塞	

事务 A 获取到(5, 10]区间的间隙锁，不允许其他的 DDL 操作，在事务提交、间隙锁释放之前，事务 B 也获取到了间隙锁(5, 10]，这时两个事务因互不相让就处于死锁状态。

> **提　示**
>
> 如果把事务的隔离级别降级为提交读，则间隙锁会自动失效。

14.3.9　记录锁

记录锁强制锁定索引记录（作用于唯一索引）。如果执行"select * from userinfo where id=4 for update;"这条语句，就会在 id=4 的索引记录上加锁，以阻止其他事务插入、更新、删除 id=4 的这一行数据。也就是说，当一个会话执行这条语句之后，其他会话执行下面这几种语句都会处于阻塞状态：

```
select * from userinfo where id=1 for update;
delete from userinfo where id=4;
insert into userinfo values(4,'hh',18);
```

阻塞时间过长可能会抛出如下异常：

```
ERROR 1205 (HY000): Lock wait timeout exceeded; try restarting transaction
```

14.3.10　临键锁

临键锁作用于非唯一索引，是记录锁与间隙锁的组合。

例 14.9　临键锁示例，数据如表 14-4 所示。

表14-4　提供的数据

id（主键）	Name	age
5	name1	18
10	name2	19
15	name3	20
20	name4	21
25	name5	22

临键锁的封锁范围既包含索引记录，又包含索引之前的区间，即(–∞,5]、(5,10]、(10,15]、(15,20]、(20,25]、(25,+∞)。

在事务中执行如下语句：

```
Update userinfo set age=19 where id= 10;
Select * from userinfo where id=10 FOR update;
```

这两个语句都会锁定(5,10]、(10,15]这两个区间，即 InnoDB 会获取该记录行的临键锁，并同时获取该记录行下一个区间的间隙锁。

14.3.11 死锁

死锁是指两个或两个以上的事务在执行过程中因争抢锁资源而造成的互相等待的现象。死锁的操作流程如图 14-3 所示。

图 14-3　死锁的操作流程

图中的两种情况即为死锁产生的常规情景。事务 A 等着事务 B 释放锁，事务 B 等着事务 A 释放锁，就会出现两个事务相互等待且一直等待下去的情况。

避免死锁的方法有两种：第一种是等待事务超时主动回滚；第二种是进行死锁检查，主动回滚某条事务，让别的事务能继续走下去。相关命令如下：

```
# 查看正在被锁的事务
select * from information_schema.innodb_trx;
# 通过如下命令杀死当前事务进程来释放锁
kill trx_id;
```

14.4　事务隔离级别

我们已经知道事务有隔离性，而事务的隔离性又分为 4 大隔离级别。

（1）未提交读：事务 A 对数据库进行修改，但未提交。此时开启事务 B，在事务 B 中能读到事务 A 对数据库进行的未提交数据的修改。

（2）提交读：事务 A 对数据库进行修改，但未提交。此时开启事务 B，在事务 B 中不能读到事务 A 对数据库的修改。在事务 B 还没有关闭时，事务 A 提交对数据库的修改，这时，在事务 B 中可以查到事务 A 对数据库的修改。不过这种提交读存在一个问题，即在同一个事务中对数据库查询两次，两次的结果不一样。

（3）可重复读：在同一个事务内对于同一条 SQL 语句在不同时刻查询返回的结果一致，这被称为可重复读，可重复读是 InnoDB 的默认级别。在 SQL 标准中，该隔离级别消除了不可重复读，但是还存在幻读。

（4）串行化：在开启事务 A 时会产生表级锁，此时别的事务会等待事务 A 结束后才会开启。一般数据库都不会采用串行化，因为无论进行什么操作都不会加锁，所以不具备可用性。

在进一步了解这几种级别之前，首先需要掌握下面几个概念：

- 脏读：是指一个事务正在访问数据，并且对数据进行了修改，而这种修改还没有提交到数据库中，这时另一个事务也访问这个数据，然后使用了这个数据。
- 不可重复读（主要针对修改）：在一个事务内多次读取同一数据，当这个事务还没有结束时，另一个事务也访问该数据，那么在第一个事务的两次读数据之间，由于第二个事务的修改可能导致第一个事务两次读到的数据是不一样的，因此称为不可重复读。例如，一个编辑人员两次读取同一文档，但在两次读取之间，作者重写了该文档。当编辑人员第二次读取文档时，该文档已更改，原始读取不可重复。如果只有在作者全部完成编写后，编辑人员才可以读取文档，则可以避免该问题。
- 幻读（主要针对新增和删除）：是指当事务不是独立执行时发生的一种现象。第一个事务对一张数据表中的数据进行了修改，这种修改涉及表中的全部数据行。同时，第二个事务也修改这张表中的数据，这种修改是向表中插入了一行新数据。第一个事务的用户会发现表中还有未修改的数据行，就好像产生了幻觉一样。例如，一个编辑人员更改作者提交的文档，当生产部门将其更改内容合并到该文档的主副本时，发现作者已将未编辑的新材料添加到该文档中。如果在编辑人员和生产部门完成对原始文档的处理之前，任何人都不能将新材料添加到文档中，则可以避免该问题。

表 14-5 所示为不同的隔离级别可能会出现的问题。

表14-5 不同的隔离级别可能会出现的问题

隔离级别	脏读	不可重复读	幻读
未提交读	可能	可能	可能
提交读	不可能	可能	可能
可重复读	不可能	不可能	可能
串行化	不可能	不可能	不可能

不同的数据库采用的隔离级别也不一样。MySQL 默认的存储引擎 InnoDB 采用的是可重复读。MySQL 采用了以乐观锁为基础的 MVCC 来解决幻读。

按照锁的理论知识来说，执行"update userinfo set age=18 where id = 1;"时会给 id=1 这行数据添加一个 X 锁，当我们再执行"select * from userinfo where id = 1;"时肯定会处于阻塞状态，但是为什么实际我们能够查询到数据呢？这是因为在 InnoDB 中给每行增加了 3 个隐藏字段来实现 MVCC，其中一个字段是事务的版本号，每开启一个新事务，事务的版本号就会递增。在可重复读级别下，select 读取的创建版本号小于或等于当前事务版本号，并且删除版本号为空或大于当前事务版本号的记录，这样可以保证在读取之前记录是存在的。

当执行 insert 命令时，当前事务的版本号保存至行的创建版本号；当执行 update 命令修改一行数据时，以当前事务的版本号作为新行的创建版本号，同时将原记录行的删除版本号设置为当前事务版本号；当执行 delete 命令时，将当前事务的版本号保存至删除行的版本号。因此，当 A 事务在操作时，在这期间其他事务对 A 事务相关的数据进行任何操作（增、删、改），MySQL 都会保证 A 事务在开始事务和提交事务期间查询的数据永远是一致的。

MySQL 就是通过以上原理解决了可重复读隔离级别的幻读问题。

例 14.10 开启两个事务会话，验证可重复读隔离级别的幻读问题，示例 SQL 语句及其执行结果如下：

```
# session1 执行如下
# 开启手动提交
mysql> set autocommit =0;
Query OK, 0 rows affected (0.00 sec)
# 开启事务
mysql> begin;
Query OK, 0 rows affected (0.00 sec)
# 查询当前表数据
mysql> select * from userinfo;
+----+----------+------+
| id | name     | age  |
+----+----------+------+
|  1 | zhangsan |   18 |
|  2 | lisi     |   18 |
|  3 | dd       |   18 |
|  4 | dd       |   18 |
|  5 | aaa      |   18 |
+----+----------+------+
5 rows in set (0.00 sec)
# session2 执行如下
# 开启事务
mysql> begin;
Query OK, 0 rows affected (0.00 sec)
# 新增一条数据
mysql> insert into userinfo values(6,'cc',18);
Query OK, 1 row affected (0.00 sec)
# 删除一条数据
mysql> delete from userinfo where id=1;
Query OK, 1 row affected (0.00 sec)
# 修改一条数据
mysql> update userinfo set age=19 where id=2;
Query OK, 1 row affected (0.01 sec)
Rows matched: 1  Changed: 1  Warnings: 0
# 提交事务
mysql> commit;
Query OK, 0 rows affected (0.00 sec)

# 重新回到 session1 会话，再次查询当前 userinfo 表的数据
# 查询表数据
mysql> select * from userinfo;
+----+----------+------+
| id | name     | age  |
+----+----------+------+
|  1 | zhangsan |   18 |
|  2 | lisi     |   18 |
|  3 | dd       |   18 |
|  4 | dd       |   18 |
```

```
| 5 | aaa       | 18 |
+---+-----------+----+
5 rows in set (0.00 sec)
```

由此可见，尽管在 session1 会话查询数据时，session2 会话进行了数据的修改、删除与新增，还是保证了 session1 会话两次查询结果的数据一致性，没有出现幻读和不可重复读的问题。

例 14.11 再次开启两个事务会话，验证可重复读隔离级别的幻读问题，示例 SQL 语句及其执行结果如下：

```
# 在 session1 上执行，开启事务
mysql> begin;
# 在 session1 上执行，查询 id=2 的数据
mysql> select * from userinfo where id=2;
+----+------+------+
| id | name | age  |
+----+------+------+
|  2 | lisi |   18 |
+----+------+------+
1 row in set (0.00 sec)

# 在 session2 上执行，开启事务
mysql> begin;
# 在 session2 上执行，修改 id 等于 2 的数据，把此行数据的 age 字段值更改为 99
mysql> update userinfo set age=99 where id=2;
Query OK, 1 row affected (0.00 sec)
Rows matched: 1  Changed: 1  Warnings: 0
# 在 session2 上执行，提交事务
mysql> commit;
Query OK, 0 rows affected (0.00 sec)

# 在 session1 上执行，再次查询 id=2 的数据
mysql> select * from userinfo where id=2;
+----+------+------+
| id | name | age  |
+----+------+------+
|  2 | lisi |   18 |
+----+------+------+
1 row in set (0.00 sec)
```

如上所示，尽管在 session1 处理事务的过程中，session2 对 id 等于 2 的数据执行了修改操作，但是 session1 两次查询数据的结果是一致的。

14.5 本章练习

1. 说明你对事务的理解。
2. 说明表级锁和行级锁的区别。
3. 在数据库查询中如何利用锁机制提高数据安全性和查询数据的效率。
4. 举例说明不同的事务隔离级别的区别。

第 15 章

InnoDB 内存结构解析和优化

本章内容：

- InnoDB 的内存结构
- InnoDB 的优化配置和原理
- InnoDB 表的优化

本章将讲解 MySQL 中 InnoDB 存储引擎的内存结构和优化。作为一名高级程序员或者数据库管理员来说，InnoDB 的优化是必不可少的技能。通过学习本章的内容，读者可以根据业务需求在 InnoDB 存储引擎的基础上提高数据库的使用效率。本章从原理到具体的配置都给出了大量示例，以方便读者理解。

15.1　InnoDB 内存结构解析和优化实战

本节将学习 InnoDB 架构图中的内存结构模型。InnoDB 架构图如图 15-1 所示。

图 15-1　InnoDB 架构图

15.1.1　缓冲池原理和优化

缓冲池是主内存中的一个区域，在 InnoDB 访问缓存表和索引数据时使用。缓冲池中存储的是用户经常使用的数据，从而可以加快处理速度。在专用服务器上，多达 80%的物理内存分配给缓冲池。为了提高大量读取操作的效率，缓冲池被划分为多个页面。为了提高缓存管理的效率，缓冲池以页面链表的方式来实现，对于很少使用的数据，利用最近最少使用算法（Least Recently Used，LRU）的变体将它从缓存中删除。下面我们来了解 LRU 算法。

缓冲池使用 LRU 算法的变体来管理页面的列表。该算法的核心思想是维护一个页面列表，其中每个页面都有一个关联的时间戳；如果数据最近被访问过，那么将来被访问的概率也很高。当需要向缓冲池添加新页面时，最近最少使用的页面会被逐出，并将新页面添加到列表中间。这意味着，随着时间的推移，页面在列表中的位置会不断变化，而最近最少使用的页面则会被移动到列表的末尾。为了实现 LRU 算法，通常使用两个子列表来表示缓冲池中的页面列表。缓冲池列表如图 15-2 所示。

第 15 章 InnoDB 内存结构解析和优化

图 15-2　缓冲池列表

在列表头部存放的是最近访问的新页面的子列表，在尾部存放的是最近访问的旧页面的子列表。该算法将经常使用的页面保留在新的子列表中。旧的子列表包含不太常用的页面，这些页面是要驱逐的候选页面。默认情况下，算法操作如下：

（1）缓冲池的 3/8 专用于存放旧子列表。

（2）列表的中点是新子列表的尾部与旧子列表的头部相连接的边界。

（3）当 InnoDB 将页面读入缓冲池时，最初将它插入中点（旧子列表的头部），这些读取的操作是由用户发起的操作（例如 SQL 查询）或者由 InnoDB 自动执行的预读操作。

（4）若访问了旧子列表中的页面，则将它移动到新子列表头部的操作会根据不同的情况而触发不同的操作：如果页面是由于用户操作需要而被读取的，则第一次访问会立即触发，即页面会变得"年轻"；如果页面是由于预读操作而被读取的，则第一次访问不会立即触发，并且在页面被逐出之前也根本不会触发。

（5）随着数据库的运行，缓冲池中未被访问的页面会通过向列表尾部移动来"老化"，即新子列表中的页面会随着其他页面的更新而老化，旧子列表中的页面也会随着新页面不断插入中点而老化。最终，一个未使用的页面因到达旧子列表的尾部而被驱逐。

默认情况下，查询读取的页面会立即移动到新的子列表中，而这些数据在缓冲池中会保持很长一段时间。比如，进行 mysqldump（数据备份）操作或全表查询等操作时，这些操作会将大量数据存储到缓冲池并驱逐等量的旧数据，即使这些新进入的数据之后不再使用。但是，如果被驱逐的数据中存在我们经常查询的数据，那么数据备份或全表查询等操作造成的大批量数据被驱逐势必对查询性能造成严重影响。

如果要利用缓冲池提高性能,那么我们可以从下面几个方面着手。

1. 调整缓冲池

1)设置缓冲池的大小

理想情况下,可以将缓冲池的大小(innodb_buffer_pool_size)设置为尽可能大的值,从而为服务器上的其他进程留出足够的内存来运行,而不会产生过多的分页。缓冲池越大,InnoDB 就越像内存数据库,从磁盘读取一次数据,然后在后续读取期间从内存访问数据。

innodb_buffer_pool_size 必须始终等于 innodb_buffer_pool_chunk_size(定义 InnoDB 缓冲池大小调整操作的块大小)×innodb_buffer_pool_instances(InnoDB 缓冲池划分的区域数)或者是其倍数。如果配置 innodb_buffer_pool_size 不等于 innodb_buffer_pool_chunk_size × innodb_buffer_pool_instance 或者不是其倍数,那么缓冲池大小将自动调整为等于 innodb_buffer_pool_ chunk_size×innodb_buffer_pool_instances 的值或其倍数值。

例 15.1　设置缓冲池的值。

innodb_buffer_pool_size 设置为 8GB,并且 innodb_buffer_pool_instances 设置为 16,innodb_buffer_pool_chunk_size 设置为 128MB(默认值)。innodb_buffer_pool_size 设置为 8GB,是一个有效值,因为它是 innodb_buffer_pool_instances(16)×innodb_buffer_pool_chunk_size(128MB)=2048MB(即 2GB)的倍数。

```
# 配置缓冲池大小(执行完需要重启数据库服务)
shell> mysqld --innodb-buffer-pool-size=8G --innodb-buffer-pool-instances=16
# 查询当前使用的线程池大小
mysql> select @@innodb_buffer_pool_size/1024/1024/1024;
+------------------------------------------+
| @@innodb_buffer_pool_size/1024/1024/1024 |
+------------------------------------------+
|                           8.000000000000 |
+------------------------------------------+
```

2)配置缓冲池块大小

如果还想配置 InnoDB 缓冲池块大小,则可以使用下面两种方式进行操作:

- 使用 mysqld 命令修改,单位是字节:

```
shell> mysqld --innodb-buffer-pool-chunk-size=134217728
```

- 修改配置文件 my.cnf:

```
[mysqld]
innodb_buffer_pool_chunk_size=134217728
```

3)查询线程池相关的默认参数值

可以根据下面的语句查询线程池相关的默认参数值。

```
# 查询 nnodb_buffer_pool_size 当前的数值
mysql> select @@innodb_buffer_pool_size;
```

```
+------------------------+
| @@innodb_buffer_pool_size |
+------------------------+
|             2147483648 |
+------------------------+
# 查询 innodb_buffer_pool_instances 当前的数值
mysql> select @@innodb_buffer_pool_instances;
+--------------------------------+
| @@innodb_buffer_pool_instances |
+--------------------------------+
|                              4 |
+--------------------------------+
# Chunk size was set to 1GB (1073741824 bytes) on startup but was
# truncated to innodb_buffer_pool_size / innodb_buffer_pool_instances

# 查询 select @@innodb_buffer_pool_chunk_size 当前的数值
mysql> select @@innodb_buffer_pool_chunk_size;
+---------------------------------+
| @@innodb_buffer_pool_chunk_size |
+---------------------------------+
|                       536870912 |
+---------------------------------+
```

4）在不重新启动服务器的情况下调整缓冲池

如果想在不重新启动服务器的情况下调整缓冲池，则可以使用下面的命令：

```
mysql> set GLOBAL innodb_buffer_pool_size=402653184;
```

在调整时，可以使用 Innodb_buffer_pool_resize_status 监控在线缓冲池大小的调整进度。

```
mysql> show STATUS where Variable_name='InnoDB_buffer_pool_resize_status';
```

2. 配置多个缓冲池

在具有足够内存的 64 位操作系统中，可以将缓冲池拆分为多个部分，以最大限度地减少并发操作之间对内存的争用。

对于缓冲池在数吉字节 GB 范围内的操作系统而言，将缓冲池划分为单独的实例，可以通过减少不同线程读取和写入缓存页面时的争用来提高并发性。在使用 innodb_buffer_pool_instances 变量配置多个缓冲池实例时，用户也可以调整 innodb_buffer_pool_size 的值。

当 InnoDB 缓冲池很大时，可以通过从内存中检索数据来满足更多的数据请求操作。在此期间可能会遇到多个线程试图同时访问缓冲池（内存争用）的问题，对于这种问题，用户可以启用多个缓冲池实例来减少内存争用。使用哈希函数将存储在缓冲池中或从缓冲池中读取的每个页面随机分配给其中一个缓冲池实例，每个缓冲池实例管理自己的空闲列表、刷新列表、LRU 算法以及连接到缓冲池的所有其他数据结构。在 MySQL 8.0 之前，每个缓冲池实例都由自己的缓冲池互斥锁保护。在 MySQL 8.0 及更高版本中，缓冲池实例互斥锁被多个列表和哈希保护互斥锁取代，以减少争用。

要启用多个缓冲池实例，可以将 innodb_buffer_pool_instances 设置为大于 1（默认值）且小于 64（最大值）的值。此变量仅在用户设置 innodb_buffer_pool_size 为 1GB 或更大时生效。用户指定的总大小在所有缓冲池之间分配。为了获得最佳效率，需要指定 innodb_buffer_pool_instances 和 innodb_buffer_pool_size 的组合，使得每个缓冲池实例至少为 1GB。

例 15.2 配置 16 个实例的线程池，示例 SQL 语句如下：

```
shell> mysqld --innodb-buffer-pool-instances=16
```

和前面的配置一样，要保证 innodb_buffer_pool_size 的值是 innodb_buffer_pool_chunk_size × innodb_buffer_pool_instances 的倍数，因此在配置前，可以根据前面提供的脚本来查看每个配置的大小，以方便配置出合理的数值范围。

3. 合理利用缓冲池

不使用严格的 LRU 算法，将不再访问的少数数据存储在缓冲池中，其目标是确保经常访问的"热"页面保留在缓冲池中，新读取的块被插入 LRU 列表的中间。默认情况下，所有新读取的页面都插入 3/8 到 LRU 列表尾部的位置。当第一次在缓冲池中访问页面时，这些页面被移到列表的前面（最近使用的一端）。因此，从未访问过的页面永远不会进入 LRU 列表的前部，并且比使用严格的 LRU 方法更快地"老化"。这种安排将 LRU 列表分成两部分，其中插入点下游的页面被认为是"旧的"，是容易被驱逐的数据。

我们可以控制 LRU 列表中的插入点，并选择 InnoDB 是否对通过表扫描或索引扫描带入缓冲池的块采用相同的优化算法。配置参数 innodb_old_blocks_pct 控制 LRU 列表中"旧"块的百分比，innodb_old_blocks_pct 的默认值是 37，对应 3/8 原固定比率，取值范围是 5（缓冲池中的新页面很快老化）～95（只有 5%的缓冲池保留给热页面，使算法接近熟悉的 LRU 策略）。

使缓冲池不被预读搅动的优化可以避免由于表扫描或索引扫描而导致的类似问题。在这些扫描中，数据页通常会被快速连续访问几次，并且再也不会被触及。配置参数 innodb_old_blocks_time 指定第一次访问页面后的时间窗口（单位为毫秒），在此期间它可以被访问而不会被移到 LRU 列表的前面（最近使用的末尾）。innodb_old_blocks_time 的默认值是 1000，增加此值会使越来越多的块更快地在缓冲池中老化。

配置操作如下：

```
# 查询 global.innodb_old_blocks_pct 参数值
select @@global.innodb_old_blocks_pct;
# 设置 global.innodb_old_blocks_pct 值
set @@global.innodb_old_blocks_pct=37;
# 查询 innodb_old_blocks_time 值(单位是毫秒)
select @@global.innodb_old_blocks_time;
# 设置 innodb_old_blocks_time 值
set @@global.innodb_old_blocks_time=1000;
```

当然，也可以直接在 my.cnf 配置文件中修改配置。

下面是针对不同业务采用的比较合理的配置策略。

（1）如果在业务中做了大量的全表扫描，就可以将 innodb_old_blocks_pct 减小，增大 innodb_old_blocks_time 的时间，不让这些无用的查询数据进入旧区域，尽量不让缓存在新区域的有用数据被立即刷掉（即遭到驱逐）。

（2）如果在业务中没有做大量的全表扫描，就可以将 innodb_old_blocks_pct 增大，减小 innodb_old_blocks_time 的时间，让有用的查询数据尽量缓存在缓冲池中，以减小磁盘 I/O，提高查询性能。

4. 缓冲池预读

在业务查询中，如果将可能会查询的页面数据异步预存到缓冲池中，将会很大程度地提高查询性能，而 InnoDB 存储引擎提供了这种预读策略。在 InnoDB 中使用两种预读算法来提高 I/O 性能，分别说明如下：

1）线性预读

根据缓冲池中按顺序访问的页面来预测可能很快需要读取哪些页面。用户可以在 InnoDB 中配置 innodb_read_ahead_threshold 参数来调整触发异步读取请求所需的顺序页面访问次数，以此来控制何时执行预读操作。配置参数 innodb_read_ahead_threshold 可以控制 InnoDB 中检查顺序页面访问模式的敏感程度。如果从一个区间中顺序读取的页数大于或等于 innodb_read_ahead_threshold 的值，则 InnoDB 启动整个后续区间的异步预读操作。innodb_read_ahead_threshold 可以设置为 0~64 的任何值，默认值为 56，值越大，访问模式检查越严格。例如，如果将该值设置为 48，则 InnoDB 仅在当前范围内的 48 个页面已被顺序访问时触发线性预读请求；如果将该值设置为 8，则即使顺序访问区中的页数少至 8，也会触发异步预读。

2）随机预读

无论这些页面的数据是否顺序读取，都会根据缓冲池中已有的页面来预测何时可能需要请求的页面。如果在缓冲池中找到来自相同范围的 13 个连续页面，则 InnoDB 异步发出请求以预读该范围的剩余页面。要启用此功能，可以把 innodb_random_read_ahead 变量设置为 No。

设置操作如下：

```
# 查询 innodb_read_ahead_threshold 参数值
select @@ innodb_read_ahead_threshold;
# 设置 innodb_read_ahead_threshold 值，也可以直接修改配置文件 my.cnf
set @@global.innodb_read_ahead_threshold=56;
# 查询 innodb_random_read_ahead 参数值(默认是 0，即 No)
select  @@innodb_random_read_ahead;
# 设置 innodb_random_read_ahead 参数值
set @@global.innodb_random_read_ahead=1;
```

5. 优化配置缓冲池刷新

配置缓冲池刷新可以控制何时执行后台刷新以及是否根据工作负载动态调整刷新速率。

InnoDB 在后台会执行一些任务，比如从缓冲池中刷新脏页。脏页是那些已被修改但尚未写入磁盘上的数据文件的页。

1）innodb_page_cleaners 变量

在 MySQL 8.0 中，缓冲池刷新由页面清理线程执行。页面清理线程的数量由 innodb_page_cleaners 变量控制，该变量的默认值为 4。但是，如果页面清理线程的数量超过缓冲池实例的数量，那么 innodb_page_cleaners 会自动设置为与 innodb_buffer_pool_instances（线程池数量）相同的值。

当脏页的百分比达到 innodb_max_dirty_pages_pct_lwm 变量定义的低百分比时，将启动缓冲池刷新。默认的低百分比是缓冲池页面的 10%，设置为 0 表示禁用低百分比预刷脏页功能。

innodb_max_dirty_pages_pct_lwm 阈值的目的是控制缓冲池中脏页的百分比，并防止脏页数量达到 innodb_max_dirty_pages_pct 变量定义的阈值，innodb_max_dirty_pages_pct_lwm 变量的默认值为 90。如果缓冲池中的脏页百分比达到 innodb_max_dirty_pages_pct 阈值，则启动刷新缓冲池。配置 innodb_max_dirty_pages_pct_lwm 时，该值应始终小于 innodb_max_dirty_pages_pct 值。如果配置 innodb_max_dirty_pages_pct_lwm 值大于 innodb_max_dirty_pages_pct 值，则会强制修改 innodb_max_dirty_pages_pct_lwm 等于 innodb_max_dirty_pages_pct。

设置操作如下：

```
# 查询 innodb_max_dirty_pages_pct_lwm 参数值
select @@innodb_max_dirty_pages_pct_lwm;
# 设置 innodb_max_dirty_pages_pct_lwm 参数值
set @@global.innodb_max_dirty_pages_pct_lwm=10;
# 查询 innodb_max_dirty_pages_pct 参数值
select @@innodb_max_dirty_pages_pct;
# 设置 innodb_max_dirty_pages_pct 参数值
set @@global.innodb_max_dirty_pages_pct=10;
```

2）其他调整缓冲池刷新行为的变量

（1）innodb_flush_neighbors 变量：

该变量定义从缓冲池中刷新页面是否会在同一范围内刷新其他脏页。默认设置为 0，表示禁用 innodb_flush_neighbors，相同范围内的脏页不会被刷新，对于 SSD 设备而言，建议使用此设置。将值设置为 1，表示会刷新相同范围内的连续脏页。将值设置为 2，表示以相同的范围刷新脏页。当数据表的数据存储在传统 HDD 存储设备上时，与在不同时间刷新单个页面相比，在一次操作中刷新相邻页面会减少 I/O 开销（主要用于磁盘查找操作），因此设置 innodb_flush_neighbors 的值为 2。

（2）innodb_lru_scan_depth 变量：

该变量控制 LRU 列表中可用页的数量，页面清理器线程会定期扫描并查找要刷新的脏页。这个操作是由页面清理器线程在后台每秒执行一次。如果在工作负载下有空闲的 I/O 容量，则可以考虑增加该值。相反，如果写入密集型的工作负载导致 I/O 容量饱和，尤其是在缓冲池较大的情况下，可以减小该值。

> **提 示**
>
> innodb_flush_neighbors 和 innodb_lru_scan_depth 变量主要用于写入密集型的工作负载。对于大量 DML 活动，如果刷新不够积极，则刷新可能会拖后；如果刷新太积极，则磁盘写入可能会使 I/O 容量饱和。理想的设置取决于工作负载、数据访问模式和存储配置（例如，数据存储在 HDD 还是 SSD 设备上）。

设置操作如下：

```
# 查询 innodb_flush_neighbors 参数值
select  @@innodb_flush_neighbors;
# 设置 innodb_flush_neighbors 参数值
set @@global.innodb_flush_neighbors=0;
# 查询 innodb_lru_scan_depth 参数值
select  @@innodb_lru_scan_depth;
# 设置 innodb_lru_scan_depth 参数值
set @@global.innodb_lru_scan_depth=1024;
```

6. 缓冲池预热

为了减少重新启动服务器后的预热时间，InnoDB 在服务器关闭时为每个缓冲池保存一些最近使用的数据页面，并在服务器启动时恢复这些页面。存储的最近使用页面将由 innodb_buffer_pool_dump_pct 变量定义。

重新启动繁忙的服务器后，通常会有一个吞吐量增加的预热期，因为缓冲池中的磁盘页面被加载回内存。我们可以通过预加载缓冲池中的磁盘页面来缩短预热时间。此外，页面的加载在后台进行，不会延迟数据库的启动。除了在关闭时保存缓冲池状态并在启动时恢复它之外，还可以在服务器运行时随时保存和恢复缓冲池状态。例如，可以在稳定的工作负载下达到稳定的吞吐量后保存缓冲池状态。

1）配置缓冲池页面转储的百分比

在将页面从缓冲池转储出去之前，可以通过设置 innodb_buffer_pool_dump_pct 变量来配置要转储的最近使用的缓冲池页面的百分比。如果计划在服务器运行时转储缓冲池页面，则可以通过动态配置该变量来实现，示例 SQL 语句如下：

```
set global innodb_buffer_pool_dump_pct=40;
```

如果计划在服务器关闭时转储缓冲池页面，则可以在配置文件 my.cnf 中进行设置（默认值是 25，表示转储最近常用的 25% 的页面），配置操作如下：

```
[mysqld]
innodb_buffer_pool_dump_pct=40
```

2）在关机时保存缓冲池状态并在启动时恢复它

如果要在服务器关闭时保存缓冲池的状态，则需要在关闭服务器之前执行以下 SQL 语句（默认情况下启用）：

```
set global innodb_buffer_pool_dump_at_shutdown=ON;
```

如果要在服务器启动时恢复缓冲池状态，则需要在启动服务器时执行以下 SQL 语句（默认情况下启用）：

```
mysqld --innodb-buffer-pool-load-at-startup=ON;
```

3）在线保存和恢复缓冲池状态

如果要在服务器运行时保存缓冲池的状态，则执行以下 SQL 语句：

```
set global innodb_buffer_pool_dump_now=on;
```

如果要在服务器运行时恢复缓冲池状态，则执行以下 SQL 语句：

```
set global innodb_buffer_pool_load_now=on;
```

4）显示缓冲池转储进度

如果要在将缓冲池状态保存到磁盘时显示进度，则执行以下 SQL 语句：

```
show status like 'innodb_buffer_pool_dump_status';
```

如果操作尚未开始，则返回 not started。如果操作已经完成，则打印完成时间。如果操作正在进行，则提供状态信息（例如 Dumping buffer pool 2/7 等）。

5）显示缓冲池加载进度

如果要在加载缓冲池时显示进度，则执行以下 SQL 语句：

```
show status like 'innodb_buffer_pool_load_status';
```

如果操作尚未开始，则返回 not started。如果操作已经完成，则打印完成时间（例如 Finished at 110505 12:23:24）。如果操作正在进行，则提供状态信息（例如已加载 123/22301 页）。

6）中止缓冲池加载操作

如果要中止缓冲池加载操作，则执行以下 SQL 语句：

```
set global innodb_buffer_pool_load_abort=on;
```

如果想要了解缓冲池的加载进度，则可以使用性能模式监控缓冲池加载进度，具体操作如下：

步骤 01 启用 stage/innodb/buffer pool load 仪器：

```
mysql> update performance_schema.setup_instruments set enabled = 'yes'
    ->        where name like 'stage/innodb/buffer%';
Query OK, 0 rows affected (0.00 sec)
```

```
Rows matched: 1  Changed: 0  Warnings: 0
```

步骤 02 启用活动事件 events_stages_current、events_stages_history 和 events_stages_history_long：

```
mysql> update performance_schema.setup_consumers set enabled = 'yes'
    ->        where name like '%stages%';
Query OK, 3 rows affected (0.00 sec)
Rows matched: 3  Changed: 3  Warnings: 0
```

步骤 03 通过启用 innodb_buffer_pool_dump_now 变量转储当前缓冲池状态：

```
mysql> set global innodb_buffer_pool_dump_now=on;
Query OK, 0 rows affected (0.00 sec)
```

步骤 04 检查缓冲池转储状态以确保操作已完成：

```
mysql> show status like 'innodb_buffer_pool_dump_status'\G;
*************************** 1. row ***************************
Variable_name: Innodb_buffer_pool_dump_status
        Value: Buffer pool(s) dump completed at 210608 23:45:31
1 row in set (0.00 sec)
```

步骤 05 通过启用 innodb_buffer_pool_load_now 变量加载缓冲池：

```
mysql> set GLOBAL innodb_buffer_pool_load_now=ON;
Query OK, 0 rows affected (0.00 sec)
```

步骤 06 查询 performance_schema.events_stages_current 表来检查缓冲池加载操作的当前状态。work_completed 列显示加载的缓冲池页面数，work_estimated 列表示剩余工作的估计值（以页为单位）。

```
mysql> select event_name, work_completed, work_estimated
    ->        from performance_schema.events_stages_current;
Empty set (0.00 sec)
```

7. InnoDB 相关监控指标

InnoDB 可以使用 show engine innodb status 输出有关缓冲池操作的指标，操作如下：

```
mysql> show engine innodb status\G
*************************** 1. row ***************************
  Type: InnoDB
  Name: 
Status: 
=====================================
2021-06-09 00:04:59 139918629566208 INNODB MONITOR OUTPUT
...
# 缓冲池相关的指标
BUFFER POOL AND MEMORY
----------------------
Total large memory allocated 137035776
Dictionary memory allocated 464915
Buffer pool size   8192
Free buffers       7128
Database pages     1060
Old database pages 411
Modified db pages  0
```

```
Pending reads      0
Pending writes: LRU 0, flush list 0, single page 0
Pages made young 2, not young 0
0.00 youngs/s, 0.00 non-youngs/s
Pages read 883, created 177, written 610
0.00 reads/s, 0.00 creates/s, 0.00 writes/s
No buffer pool page gets since the last printout
Pages read ahead 0.00/s, evicted without access 0.00/s, Random read ahead 0.00/s
LRU len: 1060, unzip_LRU len: 0
I/O sum[0]:cur[0], unzip sum[0]:cur[0]
--------------
ROW OPERATIONS
--------------
0 queries inside InnoDB, 0 queries in queue
0 read views open inside InnoDB
Process ID=1, Main thread ID=139918270789376 , state=sleeping
Number of rows inserted 9, updated 2, deleted 2, read 75
0.00 inserts/s, 0.00 updates/s, 0.00 deletes/s, 0.00 reads/s
Number of system rows inserted 84, updated 364, deleted 19, read 5248
0.00 inserts/s, 0.00 updates/s, 0.00 deletes/s, 0.00 reads/s
----------------------------
END OF INNODB MONITOR OUTPUT
============================

1 row in set (0.00 sec)
```

表 15-1 所示是相关指标的说明。

表 15-1 相关指标的说明

指标名称	备注	说明
Total memory allocated	分配的总内存	为缓冲池分配的总内存（以字节为单位）
Dictionary memory allocated	分配的字典内存	为 InnoDB 数据字典分配的总内存（以字节为单位）
Buffer pool size	缓冲池大小	分配给缓冲池的总页面大小
Free buffers	空闲缓冲区	缓冲池空闲列表的总大小（以页为单位）
Database pages	数据库页面	缓冲池 LRU 列表的总大小（以页为单位）
Old database pages	旧的数据库页面	缓冲池旧 LRU 子列表的总大小（以页为单位）
Modified db pages	修改的数据库页面	当前在缓冲池中修改的页面数
Pending reads	等待读入的页面	等待读入缓冲池的缓冲池页面数
Pending writes LRU	等待写入的旧脏页数	等待从 LRU 列表写入缓冲池的旧脏页数
Pending writes flush list	等待刷新的缓冲池页面数	检查点期间要刷新的缓冲池页面数
Pending writes single page	等待写入的单页	等待写入缓冲池的单页面
Pages made young	年轻页面总数	缓冲池 LRU 列表中年轻的页面总数
Pages made not young	老年页面总数	缓冲池 LRU 列表中未变年轻的页面总数（保留在旧子列表中但未变年轻的页面）
youngs/s	年轻页面的平均访问次数	年轻页面每秒平均的访问次数

(续表)

指标名称	备注	说明
non-youngs/s	老年页面的平均访问次数	老年页面每秒平均的访问次数
Pages read	读取的页面	从缓冲池读取的总页面数
Pages created	创建的页面	在缓冲池中创建的总页面数
Pages written	写入的页面数	写入缓冲池的总页面数
reads/s	读取/秒	每秒从缓冲池页面执行读取的平均次数
creates/s	创建/秒	每秒在缓冲池页面执行创建的平均次数
writes/s	写入/秒	每秒在缓冲池页面执行写入的平均次数
Buffer pool hit rate	缓冲池命中率	从缓冲池读取的页面与从磁盘存储读取的缓冲池页面命中率
Pages read ahead	预读页面	预读操作每秒的平均值
Pages evicted without access	未被访问而被驱逐的页面数	未被访问而从缓冲池中驱逐的页面的每秒平均值
Random read ahead	随机预读	随机预读操作的每秒平均值

表 15-1 中有两个指标需要特别注意：

（1）youngs/s 指标仅适用于年轻页面，它基于页面访问次数。对给定页面的多次访问都被计算在内。如果 youngs/s 在没有发生大扫描时是非常低的值，则考虑减少延迟时间或增加用于旧子列表的缓冲池的百分比。增加百分比会使旧的子列表变大，因此该子列表中的页面移动到尾部需要更长的时间，这增加了再次访问这些页面并使其变年轻的可能性。

（2）non-youngs/s 指标仅适用于旧页面，它基于页面访问次数。对给定页面的多次访问都被计算在内。如果 non-youngs/s 在执行大型表扫描时没有看到更高的值，则增加延迟值。

15.1.2 变更缓冲区原理与优化

1. 变更缓冲区的原理

变更缓冲区是一种特殊的数据结构，当辅助索引页不在缓冲池中时，它会缓存这些页的更改数据。缓存的更改可能由插入、更新或删除操作导致，将在以后页面通过其他读取操作加载到缓冲池时合并这些更改数据。

当我们进行插入、更新、删除操作时，可能会导致修改二级索引。如果受影响的索引页不在缓冲池中或者二级索引页不在缓冲池中，则可以利用变更缓冲区对二级索引的修改进行缓存，这样可以避免昂贵的随机访问 I/O 操作，即插入、更新、删除操作。这些操作可能需要立即从磁盘中读取受影响的索引页。当页面被其他读取操作读取到缓冲池中时，变更缓冲区可以稍后分批进行，从而提高性能。

要变更缓冲区，可以修改 innodb_change_buffering，innodb_change_bufferingde 的参数说明

如下：

- all: 默认值，缓存插入、删除和清除操作。
- none: 不缓存任何操作。
- inserts: 缓存插入操作。
- deletes: 缓存删除标记操作。
- changes: 缓存插入和删除标记操作。
- purges: 缓存在后台发生的物理删除操作。

变更缓冲区的操作如下：

```
# 修改 innodb_change_buffering 参数值
mysql> set global innodb_change_buffering=all;
Query OK, 0 rows affected (0.00 sec)
# 查询 innodb_change_buffering 参数
mysql> select @@innodb_change_buffering;
+---------------------------+
| @@innodb_change_buffering |
+---------------------------+
| all                       |
+---------------------------+
1 row in set (0.00 sec)
```

当然，也可以修改配置文件 my.cnf 中的相关配置。

2. 变更缓冲区的优化

innodb_change_buffer_max_size 变量允许将更改缓冲区的最大容量配置为缓冲池总容量的百分比。默认情况下，innodb_change_buffer_max_size 设置为 25，最大值也可以设置为 50。

在具有大量插入、更新和删除活动的 MySQL 服务器上，建议增大 innodb_change_buffer_max_size。如果在 MySQL 服务器上因为存储静态数据或者更改缓冲区消耗了共享缓冲池太多的内存空间，从而导致页面比期望的更快地退出缓冲池，在这种情况下建议减小 innodb_change_buffer_max_size 值。innodb_change_buffer_max_size 变量的设置是动态的，它允许修改设置而无须重新启动服务器。

变更缓冲区的操作如下：

```
# 查询当前的变更缓冲区大小（即容量）
mysql> select @@innodb_change_buffer_max_size;
+---------------------------------+
| @@innodb_change_buffer_max_size |
+---------------------------------+
|                              25 |
+---------------------------------+
1 row in set (0.00 sec)
# 修改变更缓冲区
mysql> set global innodb_change_buffer_max_size=30;
Query OK, 0 rows affected (0.00 sec)
mysql> select @@innodb_change_buffer_max_size;
```

```
+--------------------------------+
| @@innodb_change_buffer_max_size |
+--------------------------------+
|                             30 |
+--------------------------------+
1 row in set (0.00 sec)
```

3. 变更缓冲区的注意事项

变更缓冲区需要注意如下事项：

（1）变更缓冲区仅支持二级索引，不支持聚簇索引、全文索引和空间索引。全文索引有其自己的缓存机制。

（2）如果索引包含降序索引列或主键包含降序索引列，则辅助索引不支持更改缓冲。

（3）若在页面可用之前将页面读入缓冲池，则有关缓冲的变更会在读取完成后合并。

（4）变更缓冲区合并作为后台任务执行。

（5）变更缓冲区是持久性的，可以在系统崩溃后恢复。重新启动后，更改缓冲区合并操作将作为正常操作的一部分进行恢复。

（6）变更的页面将采用与刷新占用缓冲池其他页面相同的刷新机制进行刷新。

15.1.3 自适应哈希索引原理与配置

自适应哈希索引能够使 InnoDB 在具有适当的工作负载和足够的缓冲池内存的系统上执行操作，更像是内存中的数据库操作，且不会牺牲事务功能或可靠性。通过 innodb_adaptive_hash_index 变量启用自适应哈希索引。

自适应哈希索引是根据观察到的搜索模式，使用索引键的前缀来构建的。前缀可以是任意长度，也可能只有 B_Tree 中的某些值出现在哈希索引中。哈希索引是针对经常访问的索引页面按需构建的。如果一个数据表几乎完全适合存储在内存，那么哈希索引通过启用任何元素的直接查找来加速查询，将索引值转换为某种指针。InnoDB 具有监视索引搜索的机制，如果 InnoDB 注意到查询可以从构建哈希索引中受益，则将会自动构建哈希索引。

哈希索引的操作如下：

```
# 设置 innodb_adaptive_hash_index（1 表示开启）
mysql> set global innodb_adaptive_hash_index=1;
Query OK, 0 rows affected (0.00 sec)
# 查询 innodb_adaptive_hash_index 变量的值
mysql> select @@innodb_adaptive_hash_index;
+------------------------------+
| @@innodb_adaptive_hash_index |
+------------------------------+
|                            1 |
+------------------------------+
1 row in set (0.00 sec)
```

15.1.4 日志缓冲区

日志缓冲区是保存要写入磁盘上的日志文件的数据的内存区域。日志缓冲区大小由 innodb_log_buffer_size 变量定义，默认大小为 16MB。日志缓冲区的内容会定期刷新到磁盘。大型日志缓冲区使大型事务能够运行，而无须在事务提交之前重做日志数据写入磁盘。因此，如果有更新、插入或删除许多行的事务，则增加日志缓冲区的大小以节省磁盘 I/O。innodb_flush_log_at_trx_commit 变量控制日志缓冲区的内容如何写入和刷新到磁盘。innodb_flush_log_at_timeout 变量控制日志刷新频率。

设置日志缓冲区的操作如下：

```
# 查询 select@@innodb_log_buffer_size 变量的值
mysql> select @@innodb_log_buffer_size;
+--------------------------+
| @@innodb_log_buffer_size |
+--------------------------+
|                 16777216 |
+--------------------------+
1 row in set (0.00 sec)

# select @@innodb_log_buffer_size 变量值是只读的，所以要在配置文件 my.cnf 中修改配置
mysql> set global innodb_log_buffer_size=16777216;
ERROR 1238 (HY000): Variable 'innodb_log_buffer_size' is a read only variable

# 查看 innodb_flush_log_at_trx_commit 变量的值
mysql> select @@innodb_flush_log_at_trx_commit;
+----------------------------------+
| @@innodb_flush_log_at_trx_commit |
+----------------------------------+
|                                1 |
+----------------------------------+
1 row in set (0.00 sec)
# 修改 innodb_flush_log_at_trx_commit 变量的值
mysql> set GLOBAL innodb_flush_log_at_trx_commit=1;
Query OK, 0 rows affected (0.00 sec)
```

15.2　InnoDB 表实战调优

InnoDB 是 MySQL 客户端在可靠性和并发性都很重要的生产数据库中使用的存储引擎。本节介绍如何优化 InnoDB 表的数据库操作。

15.2.1　优化 InnoDB 表的存储

在使用 MySQL 时，可能会发现尽管一张表删除了许多数据，但是这张表的数据文件和索引文件却没有变小，这是因为 MySQL 在删除数据（特别是有 text 和 blob 列的表）时会留下许多数据碎片，这些碎片会占据数据的存储空间，所以数据文件和索引文件的大小没有改变。

这些碎片在以后插入数据时可能会被再度利用起来，当然也有可能一直存在。这些碎片不仅额外增加了存储代价，同时也因为数据碎片化降低了表的扫描效率。

如果已经删除了表的一大部分数据，或者已经对含有可变长度行的表（含有 varchar、blob 或 text 列的表）进行了很多更改，结果发现数据文件尺寸并没有减小，这是由于在执行删除操作后在数据文件中留下碎片所致，则可以使用 optimize table 来重新利用未使用的空间，并整理数据文件的碎片。

optimize table 操作会复制数据表的部分数据并重建索引。重建索引改进了索引内的数据，并且减少了表空间和磁盘上的碎片，至于能够整理出多少碎片，则会因每张表中的数据而异。如果表很大或者正在重建的索引不适合缓冲池，则此操作可能会很慢。另外，向表中添加大量数据后，第一次运行 optimize table 操作通常比以后运行慢得多。

optimize table 只对 MyISAM 和 BDB 存储引擎的数据表起作用。

下面是相关的操作命令：

- 如果是 MyISAM 存储引擎，则执行如下 SQL 语句：

```
optimize table tablename;
```

- 如果是 BDB 存储引擎，则执行如下 SQL 语句：

```
alter table tablename engine='innodb';
```

例 15.3 在 InnoDB 的表上执行语句 optimize table tablename 会报错，示例 SQL 语句及其执行结果如下：

```
mysql> optimize table userinfo \G;
*************************** 1. row ***************************
   Table: demo.userinfo
      Op: optimize
Msg_type: note
Msg_text: Table does not support optimize, doing recreate + analyze instead
*************************** 2. row ***************************
   Table: demo.userinfo
      Op: optimize
Msg_type: status
Msg_text: OK
2 rows in set (0.01 sec)
```

如上所示，直接在 InnoDB 的表中执行语句 optimize table tablename 会报错。

如果想要查询具体释放了多少存储空间，可以执行下面的命令：

```
show table status from demo like 'userinfo' \G;
```

例 15.4 查询 demo 数据库中 userinfo 表当前的存储情况，示例 SQL 语句及其执行结果如下：

```
mysql> show table status from demo like 'userinfo' \G;
*************************** 1. row ***************************
          Name: userinfo
        Engine: InnoDB
```

```
        Version: 10
     Row_format: Dynamic
           Rows: 2
 Avg_row_length: 8192
    Data_length: 16384
Max_data_length: 0
   Index_length: 0
      Data_free: 0
 Auto_increment: 3
    Create_time: 2021-06-12 01:07:15
    Update_time: NULL
     Check_time: NULL
      Collation: utf8mb4_0900_ai_ci
       Checksum: NULL
 Create_options:
        Comment:
1 row in set (0.00 sec)
```

从上面的结果可以看出当前数据表占用的具体空间（Data_free），可以在优化表前后执行此语句来计算优化的空间。

15.2.2　InnoDB 事务处理优化

要优化 InnoDB 事务处理，需要在事务功能的性能开销和服务器的工作负载之间找到理想的平衡点。比如，一个应用每秒提交上千次或者一个应用 2～3 小时提交一次，都会遇到性能问题。

默认情况下，MySQL 开启了自动提交（set autocommit=1），这意味着每执行一条语句都默认执行一次提交操作。如果在短时间内进行多次提交，将会消耗大量的性能。为了提高性能，我们可以将多个相关的数据更改操作封装到单个事务中。具体操作方式有两种：

（1）关闭自动提交（set autocommit=0），完成多次操作之后，再执行commit进行批量提交。

（2）开启事务，执行start transaction语句，多次操作完成之后，再执行commit进行批量提交。

15.2.3　InnoDB 只读事务优化

InnoDB 可以避免为已知只读的事务设置事务 id 相关的开销，只有可能执行写操作或锁定读取的事务才需要事务 id。因为只读事务不存在数据的修改与删除，所以数据库会对只读事务进行一些优化，比如不启动回滚段，不记录回滚日志。

只读事务操作规则如下：

（1）如果一次执行单条查询语句，则没有必要启用事务支持，数据库默认支持 SQL 执行期间的读一致性（set autocommit=1），除非是有了 select * from table for update 等语句。

（2）如果一次执行多条查询语句，比如统计查询、报表查询，则在这种场景下，多条查询 SQL 必须保证整体的读一致性，若在前一条 SQL 语句查询之后与后一条 SQL 语句查询之前，数据被其他用户修改了，则这次整体的统计查询将会出现读数据不一致的状态，此时应该启用只读事务支持。具体操作如下：

```
start transaction read only;
 ---操作语句
commit;
```

15.2.4 优化 InnoDB 重做日志

1. MySQL 中一条 SQL 语句的执行过程详解

想要优化重做日志（Redo Log），首先要了解执行一条 SQL 语句时，InnoDB 存储引擎到底做了什么？如图 15-3 所示是执行一条修改语句时，InnoDB 存储引擎和 MySQL 服务器各自参与处理的业务流程图。

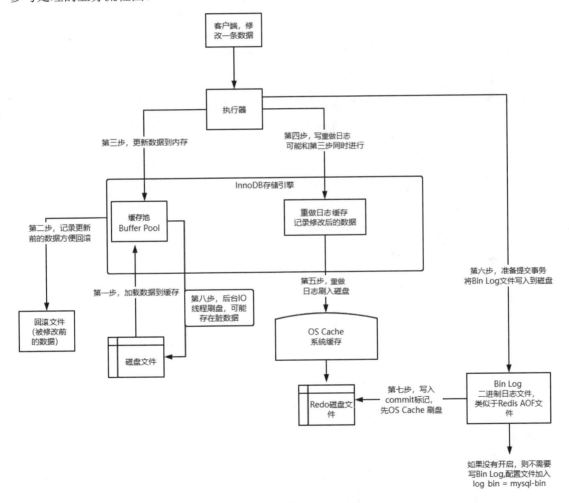

图 15-3 一条修改语句的执行流程图

下面将详细介绍相关的名词和整个执行流程。

- 缓冲池：是 MySQL 一个非常重要的组件，因为针对数据库的增、删、改操作都是在缓冲池中完成的。
- 回滚日志：记录的是数据操作前的状态，用于在需要时回滚至被修改前的数据。
- 重做日志：记录的是数据被操作后的状态（重做日志是 InnoDB 存储引擎特有的）。
- 二进制日志：记录了整个操作（属于 MySQL 服务自身的日志）。
- 日志缓冲区：其中未刷到磁盘的日志称为脏日志。

整个执行流程如下：

（1）客户端发送语句到 MySQL 数据库，将根据 MySQL 的执行计划来查询数据：先从缓冲池中查询数据，如果没有，就去数据库中查询，如果查询到了，就将它放到缓冲池中（第一步）。

（2）数据被缓存到缓冲池的同时，会写入回滚日志文件（第二步）。

（3）更新操作会在缓冲池中完成，将更新后的数据添加到重做日志缓冲区中以后就可以提交事务了（第三步），而提交的同时会做以下 3 件事：

① 将重做日志缓冲区中的数据刷入重做日志文件中（第四步和第五步）。

② 将本次操作记录写入二进制日志文件中（第六步）。

③ 将二进制日志文件名和更新内容在二进制日志文件中所在的位置都记录到重做日志文件中，同时在重做日志最后添加 commit 标记。如果在 my.cnf 中没有开启二进制日志，则直接对重做日志添加 commit 标记（第七步）。

对于将重做日志缓冲区的数据写入 Redo 磁盘文件中的策略，可以通过 innodb_flush_log_at_trx_commit 参数来设置（需要在 my.cnf 中配置）。

innodb_flush_log_at_trx_commit 不同参数的含义说明如下：

- 0：事务提交时不会将日志缓冲区中的日志立即写入系统缓存，而是每秒写入一次系统缓存并调用同步函数 fsync() 写入磁盘文件中。也就是说，设置为 0 时是每秒刷新写入磁盘中的，当系统崩溃时会丢失一秒钟的数据。
- 1：事务每次提交都会将日志缓冲区中的日志写入系统缓存并调用同步函数 fsync() 刷新到磁盘文件中。即使系统崩溃，这种方式也不会丢失任何数据，但是因为每次提交都写入磁盘，所以 I/O 性能较差。
- 2：每次提交都仅写入系统缓存，然后每秒调用同步函数 fsync() 将系统缓存中的日志写入磁盘文件中，这种方式性能最高。

2. 重做日志优化

了解上面的整个流程之后，可以考虑使用以下准则来优化重做日志：

（1）当 InnoDB 重做日志缓冲池满了或者达到一定的阈值时，必须在检查点将缓冲池的修改内容写入磁盘。重做日志文件的大小和数量可以使用 innodb_log_file_size 和 innodb_log_files_in_group（默认为 2）参数来调节，将这两个参数相乘即可得到总的可用重做日志空间。尽管技术上并不关心是通过 innodb_log_file_size 还是 innodb_log_files_in_group 来调整重做日志空间的，不过多数情况下还是通过 innodb_log_file_size 来调节的，可以在 my.cnf 配置文件中进行配置。配置完成后重启数据库服务即可。

配置格式如下：

```
[mysqld]
innodb_log_file_size=50331648
```

也可以使用下面的 SQL 语句查询默认值：

```
mysql> show global variables like '%innodb_log%';
+------------------------------------+----------+
| Variable_name                      | Value    |
+------------------------------------+----------+
| innodb_log_buffer_size             | 16777216 |
| innodb_log_checksums               | ON       |
| innodb_log_compressed_pages        | ON       |
| innodb_log_file_size               | 50331648 |
| innodb_log_files_in_group          | 2        |
| innodb_log_group_home_dir          | ./       |
| innodb_log_spin_cpu_abs_lwm        | 80       |
| innodb_log_spin_cpu_pct_hwm        | 50       |
| innodb_log_wait_for_flush_spin_hwm | 400      |
| innodb_log_write_ahead_size        | 8192     |
| innodb_log_writer_threads          | ON       |
+------------------------------------+----------+
11 rows in set (0.00 sec)
```

由此可见，innodb_log_files_in_group 的默认值是 2，innodb_log_file_size 的默认值是 50331648（字节）。

（2）考虑增加日志缓冲区的大小。大日志缓冲区使大型事务能够运行，而无须在事务提交之前将日志写入磁盘。因此，如果有大量更新、插入或删除多行的事务，则增大日志缓冲区可以节省磁盘 I/O。日志缓冲区大小使用 innodb_log_buffer_size 变量进行配置，具体操作如下：

```
# 直接动态操作（MySQL 8.0 中可以动态配置）
set global innodb_log_buffer_size=16777216;
# 可在 my.cnf 进行配置，配置完成后重启数据库服务即可。配置格式如下
[mysqld]
global innodb_log_buffer_size=16777216
# 配置完成后，可以使用下面的语句在数据库中查询验证
show global variables like '%innodb_log%';
```

（3）配置相关 innodb_log_write_ahead_size 变量以避免"读写"，此变量定义重做日志的预写块大小。配置 innodb_log_write_ahead_size 为匹配操作系统块（OS Block）或文件系统缓存块大小。当重做日志块由于预写块大小与操作系统块或文件系统缓存块大小不匹配而未完全缓存到操作系统或文件系统时，就会发生写时读取。

为了处理重做日志块大小和操作系统块间操作数据块大小协调一致的问题，在 InnoDB 中以 512 字节一个块的方式对齐写入重做文件（ib_logfileN），而操作系统一般以 4096 字节为一个块单位读写。如果即将写入的日志块不在操作系统缓冲区缓存，就需要将对应的 4096 字

节的块读入内存,修改其中的 512 字节,然后把该块写回磁盘。引入 write-ahead 是将当前写入重做文件的偏移量整除 innodb_log_write_ahead_size 参数值,不能整除时用 0 补全,使得需要写入的内容刚好是块的倍数,然后直接覆盖写入即可,将不再需要 read-modify-write 的过程,从而提升效率。因此,innodb_log_write_ahead_size 的值最好是 4096 的倍数。

具体配置如下:

```
# 直接动态操作
set global innodb_log_write_ahead_size=8192;
# 可以在my.cnf中进行配置,配置完成后重启数据库服务即可。配置格式如下
[mysqld]
innodb_log_write_ahead_size=8192
# 配置完成后,可以使用下面的语句在数据库中查询验证
> show global variables like 'innodb_log%';
+--------------------------------+------------+
| Variable_name                  | Value      |
+--------------------------------+------------+
| innodb_log_buffer_size         | 16777216   |   # 重做日志缓冲区大小,默认16MB
| innodb_log_checksums           | ON         |   # 校验
| innodb_log_compressed_pages    | ON         |   # 是否启用压缩
| innodb_log_file_size           | 50331648   |   # 日志文件大小
| innodb_log_files_in_group      | 2          |   # 日志文件数量,默认2个日志文件
| innodb_log_group_home_dir      | ./         |   # 日志文件目录,默认datadri
| innodb_log_write_ahead_size    | 8192       |   # 表示重做日志前的块大小
+--------------------------------+------------+
```

(4) MySQL 8.0.11 引入了专用的日志写入线程,用于将重做日志记录从日志缓冲区写入系统缓冲区,并将系统缓冲区刷新到重做日志文件。以前,单独的用户线程负责这些任务,从 MySQL 8.0.22 开始,可以使用 innodb_log_writer_threads 变量启用或禁用专用日志写入线程。专用日志写入线程可以提升高并发系统的性能,但对于低并发系统,则没有必要启用专用日志写入线程,禁用专用日志写入线程可以提供更好的性能。

配置方式如下:

```
# 直接动态操作(ON:启用,OFF:禁用)
set global innodb_log_writer_threads=ON;
# 可在my.cnf中进行配置,配置完成后重启数据库服务即可。配置格式如下:
[mysqld]
innodb_log_writer_threads=ON
# 配置完成后可以使用下面的语句在数据库中查询验证
show global variables like 'innodb_log%';
```

(5) 对于以上配置的优化,用户线程不再自己去刷新重做日志,而是等待后台线程去刷新。需要注意的是,用户线程使用自旋延迟等待后台线程刷新,因此如果想要提高用户线程的性能,可以动态调整自旋延迟的高水位值和低水位值。自旋是一种通过不间断地测试来查看一个资源是否变为可用状态的等待操作,用于仅需很短的时间来等待所需资源的应用场景。优化器使用自旋延迟的方式等待重做日志的刷新,因为自旋延迟有利于减少延迟。在低并发时,减少延迟可能并没有太大效果。在高并发时,希望在自旋延迟上避免消耗处理能力,以便它可以

用于其他工作，可以设置以下变量进行优化，这些变量值定义了使用自旋延迟的边界：

- innodb_log_wait_for_flush_spin_hwm：表示最大平均日志刷新时间，超过该时间，用户线程在等待刷新重做日志时不再自旋，可以处理其他任务。默认值为 400 微秒。
- innodb_log_spin_cpu_abs_lwm：表示在等待刷新重做日志时用户线程不再自旋的最小 CPU 使用量。该值表示 CPU 内核使用量的总和。例如，默认值 80 是单个 CPU 内核的 80% 使用率。在具有多核处理器的系统上，当值为 150 时表示一个 CPU 内核的 100% 使用率加上另一个 CPU 内核的 50% 使用率。
- innodb_log_spin_cpu_pct_hwm：表示在等待刷新重做日志时用户线程不再自旋的最大 CPU 使用量。该值表示所有 CPU 内核的总处理能力的百分比，默认值为 50%。
- nnodb_log_spin_cpu_pct_hwm：表示配置变量和处理器方面的相关性。例如，如果服务器有 16 个 CPU 内核，但 mysqld 进程仅固定使用 8 个 CPU 内核，那么其他 8 个 CPU 内核将被忽略。

配置方式如下：

```
# 直接动态操作
set global  innodb_log_wait_for_flush_spin_hwm=400;
set global  innodb_log_spin_cpu_abs_lwm=80;
set global  innodb_log_spin_cpu_pct_hwm=50;
# 可以在 my.cnf 中进行配置，配置完成后重启数据库服务即可。配置格式如下
[mysqld]
innodb_log_wait_for_flush_spin_hwm=400;
innodb_log_spin_cpu_abs_lwm=80;
innodb_log_spin_cpu_pct_hwm=50;
# 配置完成后，可以使用下面的语句在数据库中查询验证
show global variables like 'innodb_log%';
```

15.2.5　InnoDB 磁盘 I/O 优化

如果已经遵循数据库设计和调优技术的最佳实践准则，但由于大量磁盘 I/O 操作导致数据库系统运行仍然很慢，就要考虑对这些磁盘 I/O 进行优化。如果利用 UNIX Top 工具或 Windows 任务管理器查看到承担数据库系统工作负载的 CPU 的使用率低于 70%，则可能是工作负载受限于磁盘，就需要进行如下优化：

1）增加缓冲池大小

当数据表的数据缓存在 InnoDB 缓冲池中时，它可以被重复查询访问，而不需要任何磁盘 I/O。使用 innodb_buffer_pool_size 变量指定缓冲池的大小。此内存区域非常重要，通常建议将 innodb_buffer_pool_size 配置为系统内存的 50%～75%。

2）调整刷盘机制

刷盘机制就是图 15-3 中第三步和第四步的业务，即把数据更新到内存及写重做日志。

默认情况下，innodb_flush_log_at_trx_commit=1 表示每次提交数据都会将日志缓冲区中的日志写入系统缓存并调用同步函数 fsync() 刷新到磁盘中。这种方式即使系统崩溃也不会丢失任何数据，但是由于每次提交都要写入磁盘，I/O 性能较差。另外，在此操作中，都是把数据

全部写入系统缓存之后再去调用同步函数进行刷新（系统缓存数据写入磁盘）。如果数据量特别大，就会导致大量磁盘操作一次性完成，性能特别低。如果想要在写系统缓存的同时进行刷新来提高效率，则需要强制从系统缓存中进行较小的、定期的数据刷新，可以使用 innodb_fsync_threshold 变量来定义一个阈值（以字节为单位）。当达到阈值时，系统缓存的内容将刷新到磁盘。innodb_fsync_threshold 变量的默认值为 0，表示仅在文件完全写入缓存后才将数据刷新到磁盘。

配置方式如下：

```
# 直接动态操作
set innodb_fsync_threshold=4;
# 可在 my.cnf 中进行配置，配置完成后重启数据库服务即可。配置格式如下
[mysqld]
innodb_fsync_threshold=4;
# 配置完成后，可以使用下面的语句在数据库中查询验证
show global variables like 'innodb_fsyn%';
```

15.2.6　配置 InnoDB 线程并发数量

InnoDB 的线程处理来自用户事务的请求，事务在提交或回滚之前可能会发出许多请求。在具有多核处理器的操作系统和服务器上，上下文切换是高效的，大多数工作负载运行良好，对并发线程数没有任何限制。在有助于最小化线程之间上下文切换的情况下，InnoDB 可以使用多种技术来限制操作系统并发执行的线程数量（也就是限制在任何时间处理的请求数量）。

当 InnoDB 收到来自用户会话的新请求时，如果并发执行的线程数超过了预先定义的线程数限制，则新请求会在再次尝试之前休眠一小段时间。休眠后无法重新调度的请求将被放入先进先出队列并最终被处理。等待锁的线程不计入并发执行的线程数，而等待的线程数量最大值是变量 innodb_concurrency_tickets 的值（默认值为 5000），如果请求进来之后，线程数量超过这个值，则无须休眠重试，直接返回错误。

变量 innodb_thread_concurrency 用来限制并发线程的数量，一旦正在执行的线程数达到此限制，则其他线程会休眠一段时间后再次尝试，休眠时间由变量 innodb_thread_sleep_delay（微秒）设置。当然，也可以直接配置变量 innodb_adaptive_max_sleep_delay 来改变运行线程数量的最大值。当前线程调度会自动向上或向下调整变量 innodb_thread_sleep_delay 的值，这种动态调整有助于线程调度机制在系统负载较轻和接近满负荷时平稳运行。

变量 innodb_thread_concurrency 的默认值是 0，表示默认情况下并行执行的线程的数量没有上限，建议设置不超过 32。

InnoDB 仅当并发线程数有限制时才导致线程休眠。当线程数没有限制时，所有线程都通过平等竞争被调度执行。也就是说，如果变量 innodb_thread_concurrency 的值是 0，则忽略变量 innodb_thread_sleep_delay 的值。当线程数有限制时（innodb_thread_concurrencyi>0），InnoDB 允许同一时间有多个请求，但是会对同时处理的线程数量进行限制，从而减少上下文切换的开

销。配置线程并发数量的操作如下：

```
# 可在 my.cnf 中进行配置，配置完成后重启数据库服务即可。配置格式如下
[mysqld]
innodb_thread_concurrency=0;
innodb_thread_sleep_delay=1000;
innodb_concurrency_tickets =1000;
# 配置完成后，可以使用下面的语句在数据库中查询验证
select @@innodb_thread_concurrency;
select @@innodb_thread_sleep_delay;
select @@innodb_concurrency_tickets;
```

15.2.7 配置 InnoDB 的 I/O 读写后台线程数量

InnoDB 使用后台线程来服务各种类型的 I/O 请求，可以使用 innodb_read_io_threads（读线程数量）与 innodb_write_io_threads（写线程数量）参数配置 I/O 读写后台线程的数量。这些参数分别表示用于读取和写入请求的后台线程数。它们在所有支持的平台上都有效。

配置这些参数的目的是使 InnoDB 在高端系统上更具可扩展性。每个后台线程最多可以处理 256 个待处理的 I/O 请求。后台 I/O 的主要来源是预读请求。InnoDB 尝试以大多数后台线程平等共享工作的方式平衡传入请求的负载。InnoDB 还尝试将来自同一范围的读取请求分配给同一线程，以增加合并请求的机会。

innodb_read_io_threads 和 innodb_write_io_threads 参数不能动态更改值，可以在 MySQL 选项文件 my.cnf 或 my.ini 中进行配置。这两个参数的默认值为 4，并且允许的值范围为 1~64。配置操作如下：

```
[mysqld]
innodb_write_io_threads=4;
innodb_read_io_threads=4;
```

15.2.8 InnoDB 的并行读线程数量

长久以来，MySQL 低版本没有提供并行查询，MySQL 在 8.0.14 版本开始有了自己的并行查询，但使用面非常窄，只适用于并行聚集索引的 count(*)，并且只是在没有 where 条件下的查询。因此，要运行并行查询需要满足如下条件：

（1）无锁查询。

（2）聚集索引。

（3）不是 insert…select。

例 15.5 同时开启 6 个线程来读取数据，示例 SQL 语句及其执行结果如下：

```
mysql> set local innodb_parallel_read_threads=6;
Query OK, 0 rows affected (0.00 sec)
mysql> select count(*) from test;
+----------+
| count(*) |
```

```
+-----------+
|   1888    |
+-----------+
1 row in set (2.98 sec)
```

在实际业务中,并行查询线程的具体数量取决于服务器的资源和数据量的大小。

15.2.9 InnoDB 的 I/O 容量

InnoDB 主线程和其他线程在后台执行的各种任务,大部分都与 I/O 相关,如从缓冲池刷新脏页,或从变更缓冲池更改相应的二级索引。InnoDB 尝试以不会对服务器的正常工作产生不利影响的方式执行这些任务,并尝试可用的 I/O 带宽以及调整其活动以使用可用的容量。innodb_io_capacity 变量定义了可用于整体 I/O 的容量,建议把该变量的值设置为系统每秒大约可执行的 I/O 操作的次数(Input/Output Operations Per Second,IOPS)。

innodb_io_capacity 变量的默认值为 200,通常大约 100 的值适用于消费级存储设备,例如高达 7200 RPM 的硬盘驱动器。更快的硬盘驱动器、RAID 配置和固态驱动器可以配置更高的值。通常,可以根据用于 InnoDB I/O 的驱动器数量来增加该值。例如,可以在使用多个磁盘或 SSD 的系统上增加该值。对于低端 SSD 来说,默认值 200 就足够了。

如果刷新任务落后,则 InnoDB 可以采用比 innodb_io_capacity 变量定义的更高的每秒 I/O 操作速率来更积极地刷新。innodb_io_capacity_max 变量定义了 InnoDB 在这种情况下由后台任务执行的最大 IOPS。

需要注意的是,如果在启动时设置了 innodb_io_capacity,但未设置 innodb_io_capacity_max,则 innodb_io_capacity_max 默认为 innodb_io_capacity 或 2000 的两倍,以较大者为准。在配置时,一般情况下,把 innodb_io_capacity_max 设置为 innodb_io_capacity 的两倍。innodb_io_capacity_max 的默认值是 2000,适用于使用固态驱动器或多个常规磁盘驱动器的工作负载。对于单个普通磁盘驱动器,建议设置在 200~400。配置方式如下:

```
# 直接动态操作,不需要重启服务
set global  innodb_io_capacity=200;
set global  innodb_io_capacity_max=2000;
# 配置完成后,可以使用下面的语句在数据库中查询验证
select @@innodb_io_capacity;
select @@innodb_io_capacity_max;
```

15.3 本章练习

1. 简单画出 InnoDB 内存结构图。
2. 写出几个 InnoDB 表优化的事项。

第 7 部分

MySQL 数据运维和读写分离架构

第 16 章　MySQL 数据的备份与还原
第 17 章　MySQL 主从架构

第 16 章

MySQL 数据的备份与还原

本章内容：

- 基准测试
- 数据备份
- 数据还原

本章将讲解 MySQL 数据库的基准测试，让用户在使用数据库时了解当前服务器的性能到底如何，对优化 MySQL 服务器提供指标。此外，本章还将详细介绍 MySQL 数据的备份、还原操作。本章通过详细的使用说明让读者实操起来更加容易。

16.1 基准测试

对数据库进行基准测试，以便了解数据库的性能情况，是非常必要的。本节将介绍 MySQL 基准测试的基本概念，以及使用 Sysbench 对 MySQL 进行基准测试的详细方法。

基准测试可以理解为针对系统的一种压力测试。基准测试不关心业务逻辑，更加简单、直接，易于测试，数据可以由工具生成，不要求真实；而压力测试一般考虑业务逻辑，要求真实的数据。

对于大多数 Web 应用来说，整个系统的瓶颈在于数据库，原因很简单：Web 应用中的其他因素（例如网络带宽、负载均衡节点、应用服务器（包括 CPU、内存、硬盘灯、连接数等）、缓存）都很容易通过增加机器水平的扩展来实现性能的提高。而对于 MySQL，由于数据一致性的要求，无法通过增加机器来分散向数据库写数据带来的压力，虽然可以通过读写分离、分库、分表来减轻压力，但是与系统其他组件的水平扩展相比，数据库仍然受到了太多的限制。

对数据库进行基准测试的作用是分析在当前配置（硬件配置、操作系统配置、数据库配置等）下，其数据库的性能表现，从而找出 MySQL 的性能阈值，并根据实际系统的要求调整配置。基准测试的指标有如下几个：

- 每秒查询数（Query Per Second，QPS）：是对一个特定的查询服务器在规定时间内处理查询数量的衡量标准，对应 fetches/sec，即每秒的响应请求数。
- 每秒处理的事务数（Transaction Per Second，TPS）：是指系统在单位时间内处理事务的数量。对于非并发的应用系统而言，TPS 与响应时间呈反比关系，实际上此时 TPS 就是响应时间的倒数。前面已经讲过，对于单用户的系统，响应时间（或者系统响应时间和应用延迟时间）可以很好地度量系统的性能，但对于并发系统而言，通常需要用 TPS 作为性能指标。
- 响应时间：包括平均响应时间、最小响应时间、最大响应时间、时间百分比等，其中时间百分比参考意义较大，如前 90%的请求的最大响应时间。
- 并发量：同时处理的查询请求的数量，即可以同时承载的正常业务功能的请求数量。

在对 MySQL 进行基准测试时，一般使用专门的工具，例如 MySQLslap、Sysbench 等。其中，Sysbench 比 MySQLslap 更通用、更强大。

16.1.1 Sysbench 介绍

Sysbench 是一个开源的、模块化的、跨平台的多线程性能测试工具，可以用来进行 CPU、内存、磁盘 I/O、线程、数据库的性能测试。目前支持的数据库有 MySQL、Oracle 和 PostgreSQL。它主要包括以下几种测试：

- CPU 性能。
- 磁盘 I/O 性能。
- 调度程序性能。
- 内存分配及传输速度。
- 数据库性能基准测试。

Sysbench 的基本语法如下：

```
sysbench [options]... [testname] [command]
```

下面说明实际使用中常用的参数和命令。

1）command 参数

在 Sysbench 中，command 参数用于指定要执行的命令类型。具体来说，有以下 3 种命令类型：

- prepare：用于生成测试数据并为后续测试做准备。
- run：用于执行实际性能测试，并输出测试结果。
- cleanup：测试完成后对数据库进行清理，以便下次测试。

命令类型可以通过命令行参数或配置文件来设置。例如，以下命令将执行 Sysbench 基准测试并输出结果：

```
sysbench cpu run
```

在上述命令中，"cpu"指示 Sysbench 执行 CPU 基准测试，"run"指示 Sysbench 执行测试运行命令。

2）config-file 参数

config-file 参数用于指定要进行测试的文件，语法如下：

```
sysbench --config-file=config oltp_point_select
```

3）options 参数

options 参数用于设置各种选项，这些选项可以影响 Sysbench 的性能和功能。以下是一些常用的 options 选项：

- mysql-host：MySQL 服务器的主机名，默认为 localhost。
- mysql-port：MySQL 服务器的端口，默认为 3306。
- mysql-user：用户名。
- mysql-password：密码。

4）MySQL 执行参数

MySQL 执行参数用于指定与 MySQL 数据库相关的各种选项。以下是一些常用的 MySQL 执行参数：

- oltp-test-mode：执行模式，包含查询、修改、删除等。
- oltp-tables-count：测试的数据表的数量。
- oltp-table-size：测试的数据表的大小。
- threads：客户端的并发连接数。

16.1.2 Sysbench 测试实操

在测试之前，需要注意下面几点：

（1）最好不要在 MySQL 服务器运行的机器上进行测试，一方面，可能无法体现网络传输速率的影响；另一方面，Sysbench 的运行会影响 MySQL 服务器的性能。

（2）可以逐步增加客户端的并发连接数，观察在连接数不同的情况下，MySQL 服务器的表现，如分别设置并发连接数为 8、16、24、32 等。

（3）如果要连续进行多次测试，则注意确保之前测试的数据已经被清理干净。

为了方便学习和演示，我们使用自定义镜像文件运行 Sysbench 来进行基准测试。测试步骤如下：

步骤 01 制作 Docker 自定义镜像。创建 Dockerfile 文件，SQL 语句如下：

```
# 创建目录
 mkdir /home/sysbench -p
# 创建文件 Dockerfile，输入下面的命令
  vi Dockerfile
# 输入 i 键，把下面的内容复制到此文件
FROM ubuntu:20.04
RUN apt update && apt install -y sysbench \
    && apt clean \
    && rm -rf /var/lib/apt/lists/* \
```

步骤 02 生成并运行自定义镜像，SQL 语句如下：

```
# 进入文件目录
cd /home/sysbench
# 生成一个名为 sysbench 的镜像
docker build -t sysbench .
# 运行镜像
docker run -i -t -d --name sysbenchtest -v /tmp/log/:/log/ sysbench
```

步骤 03 数据库准备工作。需要在进行基准测试的数据库系统中创建一个数据库 sbtest，SQL 语句如下：

```
create database sbtest
```

步骤 04 导入测试数据，SQL 语句如下：

```
# 进入 Sysbench 容器
docker exec -it sysbenchtest /bin/bash
# 导入数据
sysbench --db-driver=mysql --mysql-host=192.168.1.211
--mysql-port=3306--mysql-user=root --mysql-password=123456 --config-file=config
oltp_point_select --tables=32 --table-size=100000 prepare
```

在导入数据时，要确保连接的数据库地址、用户名和密码正确。在以上命令中，--tables=32 表示创建 32 张测试数据表，table_size=100000 表示在每张测试数据表中插入 10 万行数据，prepare 表示这是准备数据的过程。

步骤 05 进行性能测试，SQL 语句和测试结果如下：

（1）point select 测试命令如下：

```
sysbench --db-driver=mysql --mysql-host=192.168.1.211 --mysql-port=3306
--mysql-user=root --mysql-password=123456 --threads=4 --config-file=config
oltp_point_select --tables=32 --table-size=100000 run
```

测试结果如下：

```
root@4232c79473d0:/# sysbench --db-driver=mysql --mysql-host=192.168.1.211
--mysql-port=3306 --mysql-user=root --mysql-password=123456 --threads=4
--config-file=config oltp_point_select --tables=32 --table-size=100000 run
    config: No such file or directory
    sysbench 1.0.18 (using system LuaJIT 2.1.0-beta3)
    Running the test with following options:
    Number of threads: 4
    Initializing random number generator from current time
    Initializing worker threads...
    Threads started!
    SQL statistics:
        queries performed:
            read:           70982           # 执行读操作数量
            write:          0               # 执行写操作数量
            other:          0               # 执行其他操作数量
            total:          70982           # 执行操作总数量
        transactions:       70982   (7095.36 per sec.)    # 执行事务的平均速率
        queries:            70982   (7095.36 per sec.)    # 平均每秒能执行多少次查询
        ignored errors:     0       (0.00 per sec.)
        reconnects:         0       (0.00 per sec.)

    General statistics:
        total time:                         10.0027s         # 总时间
        total number of events:             70982            # 总请求数量(读、写、其他)

    Latency (ms):
            min:                                    0.15
            avg:                                    0.56
            max:                                   15.49
            95th percentile:                        0.92    # 采样计算的平均延迟
            sum:                                39955.41

    Threads fairness:
        events (avg/stddev):           17745.5000/101.23    # 采样计算的平均延迟
        Execution time (avg/stddev):       9.9889/0.00
```

（2）update index 测试命令如下：

```
    sysbench --db-driver=mysql --mysql-host=192.168.1.211 --mysql-port=3306
--mysql-user=root --mysql-password=123456 --threads=4 --config-file=config
oltp_update_index --tables=32 --table-size=100000 run
```

测试结果如下：

```
root@4232c79473d0:/# sysbench --db-driver=mysql --mysql-host=192.168.1.211
--mysql-port=3306 --mysql-user=root --mysql-password=123456 --threads=4
--config-file=config oltp_update_index --tables=32 --table-size=100000 run
    config: No such file or directory
    sysbench 1.0.18 (using system LuaJIT 2.1.0-beta3)
    Running the test with following options:
    Number of threads: 4
    Initializing random number generator from current time
    Initializing worker threads...
    Threads started!
    SQL statistics:
```

```
queries performed:
    read:                            0            # 执行读操作数量
    write:                           6801         # 执行写操作数量
    other:                           0            # 执行其他操作数量
    total:                           6801         # 执行操作总数量
transactions:                        6801  (709.36 per sec.)    # 执行事务的平均速率
queries:                             6801  (709.36 per sec.)    # 平均每秒能执行多少次查询
ignored errors:                      0     (0.00 per sec.)
reconnects:                          0     (0.00 per sec.)

General statistics:
    total time:                      10.0054s     # 总时间
    total number of events:          6801         # 总请求数量(读、写、其他)

Latency (ms):
    min:                                  1.75
    avg:                                  5.88
    max:                                  47.65
    95th percentile:                      22.04   # 采样计算的平均延迟
    sum:                                  39991.30

Threads fairness:
    events (avg/stddev):             1700.2500/7.89    # 采样计算的平均延迟
    execution time (avg/stddev):     9.9978/0.00
```

（3）read-only 测试命令如下：

```
sysbench --db-driver=mysql --mysql-host=192.168.1.211 --mysql-port=3306
--mysql-user=root --mysql-password=123456 --threads=4 --config-file=config
oltp_read_only --tables=32 --table-size=100000 run
```

测试结果如下：

```
root@4232c79473d0:/# sysbench --db-driver=mysql --mysql-host=192.168.1.211
--mysql-port=3306 --mysql-user=root --mysql-password=123456 --threads=4
--config-file=config oltp_read_only --tables=32 --table-size=100000 run
config: No such file or directory
sysbench 1.0.18 (using system LuaJIT 2.1.0-beta3)
Running the test with following options:
Number of threads: 4
Initializing random number generator from current time
Initializing worker threads...
Threads started!
SQL statistics:
    queries performed:
        read:                        48902        # 执行读操作数量
        write:                       0            # 执行写操作数量
        other:                       6986         # 执行其他操作数量
        total:                       55888        # 执行操作总数量
    transactions:                    3493   (348.85 per sec.)    # 执行事务的平均速率
    queries:                         55888  (5581.36 per sec.)   # 平均每秒能执行多少次查询
    ignored errors:                  0      (0.00 per sec.)
    reconnects:                      0      (0.00 per sec.)

General statistics:
```

```
            total time:                          10.0118s      # 总时间
            total number of events:              3493          # 总请求数量(读、写、其他)
    Latency (ms):
            min:                                 5.15
            avg:                                 11.56
            max:                                 35.49
            95th percentile:                     14.92         # 采样计算的平均延迟
            sum:                                 39955.41
    Threads fairness:
            events (avg/stddev):                 17745.5000/101.23   # 采样计算的平均延迟
            execution time (avg/stddev):         9.9889/0.00
```

以上是 3 种不同模式的简单测试。在实际测试中，我们需要注意下面几点：

（1）在开始测试之前，首先应该明确是针对整个系统的基准测试，还是针对 MySQL 的基准测试，或者二者都需要。

（2）基准测试的每一种模式都要进行多次。

（3）测试必须模拟多线程的情况，单线程不但无法模拟真实的效率，而且无法模拟阻塞甚至死锁的情况。

16.2 备　　份

数据备份是企业数据管理中极其重要的一项工作。关于数据备份、恢复，有很多应用场景及相应的方法，本节主要介绍使用 mysqldump 备份数据的方法。

mysqldump 是 MySQL 用于转存数据库的实用程序。它主要产生一个 SQL 脚本，其中包含从头重新创建数据库所必需的命令 create table insert 等。

1. 备份数据

要使用 mysqldump 导出数据，需要使用--tab 选项来指定导出文件存储的目录，该目录必须有写操作权限。

例 16.1　将数据表 userinfo 导出到/tmp 目录中。

步骤 01　修改 MySQL 配置文件，让其支持导入和导出，SQL 语句如下：

```
[mysqld]
secure_file_priv=
```

步骤 02　配置后重启 MySQL 服务。

步骤 03　执行如下语句备份 demo 数据库下的 student 数据表：

```
mysqldump -uroot -p123456 --no-create-info --tab=/tmp demo userinfo
```

步骤 04 到/tmp 目录下查看，该目录下会多出一个 userinfo.sql 文件，文件内容如下：

```
-- MySQL dump 8.23
--
-- Host: localhost    Database: demo
---------------------------------------------------------
-- Server version       3.23.58
--
-- Table structure for table `runoob_tbl`
--
create table userinfo(
   id int(10) NOT null auto_increment primary key,
   name varchar(20),
    age int(2)
);
--
-- Dumping data for table `userinfo`
--
 insert into userinfo (name,age)values ('zhangsan',26);
 insert into userinfo(name,age)values ('lisi',26);
```

2. 备份数据库

使用 mysqldump 备份数据库的语法如下：

```
mysqldump -h 服务器 -u 用户名 -p 密码 数据库名 > 备份文件.sql
```

- 单库备份

```
mysqldump -uroot -p123456 db1 > db1.sql
mysqldump -uroot -p123456 db1 table1 table2 > db1-table1-table2.sql
```

- 多库备份

```
mysqldump -uroot -p123456 --databases db1 db2 mysql db3 > db1_db2_mysql_db3.sql
```

- 备份所有库

```
mysqldump -uroot -p123456 --all-databases > all.sql
```

16.3 还 原

有数据备份就有数据还原，在 MySQL 中可以使用多种方式对备份的数据进行还原。

16.3.1 利用 source 命令导入数据库

若要使用 source 命令导入数据库，就需要先登录数据库终端。

执行 source 命令导入数据库的操作如下：

```
# 创建数据库
create database demo;
# 使用已创建的数据库
```

```
use demo;
# 设置编码
set names utf8;
# 导入备份数据库，目录就是备份后的文件目录
mysql> source /home/data/userinfo.sql
```

16.3.2 利用 load data infile 导入数据

MySQL 提供了 load data infile 语句来插入数据。

例 16.2 从当前目录中读取文件 userinfo.txt，将该文件中的数据插入当前数据库的 userinfo 数据表中，示例 SQL 语句如下：

```
Load data local infile /data/userinfo.txt' into table userinfo;
```

16.3.3 利用 mysqlimport 导入数据

mysqlimport 客户端提供了另一种命令行接口。mysqlimport 的大多数选项直接对应 load data infile 子句。

例 16.3 从文件 userinfo.txt 中将数据导入 userinfo 数据表中，示例 SQL 语句如下：

```
mysqlimport -uroot -p123456 --local userinfo userinfo.txt
```

16.4　本　章　练　习

1. 写出数据备份的语法。
2. 写出数据还原的语法。

第 17 章

MySQL 主从架构

本章内容：
- 读写分离的价值和意义
- MySQL 主从复制原理
- MySQL 主从复制环境搭建

本章将讲解读写分离的作用和意义，然后利用 MySQL 主从复制实现读写分离，最后利用 Docker 环境快速搭建 MySQL 主从复制架构，让初学者能便捷地搭建一套自己的主从复制架构。

17.1 读写分离

随着应用业务数据的不断增多，程序应用的响应速度会不断下降，在检测过程中不难发现大多数的请求都是查询操作。此时，我们可以将数据库扩展成主从复制模式，将读操作和写操作分离开来，多台数据库分摊请求，从而减少单库的访问压力，进而使应用得到优化。

读写分离的基本原理是让主数据库处理对数据的增、改、删操作，进而让从数据库处理查询操作。数据库复制用来把事务性操作导致的变更同步到集群的从数据库中。由于数据库的操作比较耗时，因此让主服务器处理写操作以及实时性要求比较高的读操作，而让从服务器处理读操作。读写分离能提高性能的原因在于主、从服务器负责各自的读和写，极大地缓解了锁的争用，其架构图如图 17-1 所示。

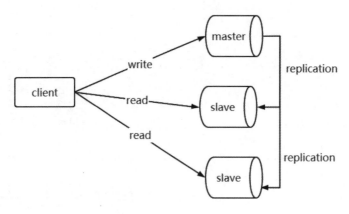

图 17-1 读写分离架构图

图 17-1 所示架构有一个主库与两个从库：主库负责写数据，从库复制读数据。随着业务发展，如果还想增加从节点来提升读性能，那么可以随时进行扩展。

17.2 MySQL 主从复制

如果需要使用 MySQL 服务器提供读写分离支持，则需要 MySQL 的一主多从架构。在一主多从的数据库体系中，多个从服务器采用异步的方式更新主数据库的变化，业务服务器执行写操作或者相关修改数据库的操作直接在主服务器上执行，读操作在各从服务器上执行。MySQL 主从复制实现原理如图 17-2 所示。

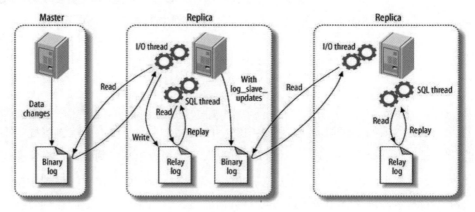

图 17-2 MySQL 主从复制实现原理

图 17-2 所示是典型的 MySQL 一主二从的架构图，其中主要涉及 3 个线程：binlog 线程、I/O 线程和 SQL 线程。每个线程说明如下：

- binlog 线程：负责将主服务器上的数据更改写入二进制日志中。
- I/O 线程：负责从主服务器上读取二进制日志，并写入从服务器的中继日志中。

- SQL 线程：负责读取中继日志并重放其中的 SQL 语句。

MySQL 服务之间数据复制的基础是二进制日志文件。一个 MySQL 数据库一旦启用二进制日志后，其作为主服务器，它的数据库中所有操作都会以"事件"的方式记录在二进制日志中，其他数据库作为从服务器，通过一个 I/O 线程与主服务器保持通信，并监控主服务器的二进制日志文件的变化。如果发现主服务器的二进制日志文件发生变化，则会把变化复制到自己的中继日志中，然后从服务器的一个 SQL 线程把相关的"事件"触发操作在自己的数据库中执行，以此实现从数据库和主数据库的一致性，也就实现了主从复制。

17.3 MySQL 主从复制架构的搭建

本节主要讲述整个 MySQL 主从复制架构的搭建过程。

MySQL 主从复制配置如下：

1）对于主服务器

（1）开启二进制日志。

（2）修改配置文件配置唯一的 server-id。

（3）获得主节点二进制日志文件名及位置。

（4）创建一个用于从节点和主节点通信的用户账户。

2）对于从服务器

（1）配置唯一的 server-id。

（2）使用主节点分配的用户账户读取主节点二进制日志。

（3）启用从节点服务。

为了方便讲解，这里使用 Docker 在一台服务器上启动主、从两个节点。如果需要两台服务器，原理相同，可以在两台服务器上分别执行如下命令：

```
# 启动主节点 ip:47.95.2.2 挂载 3306 端口
docker run -p 3306:3306 --name mysql-master -e MYSQL_ROOT_PASSWORD=123456 -d mysql:8.0
```

```
# 启动从节点 ip:47.95.2.2 挂载 3307 端口
docker run -p 3307:3306 --name mysql-slave -e MYSQL_ROOT_PASSWORD=123456 -d mysql:8.0
```

17.3.1 配置主节点

配置主节点的操作流程如下：

步骤 01 执行如下命令进入主节点容器内部，使用容器 ID 或者名称均可：

```
docker exec -it mysql-master /bin/bash
# 登录 MySQL
mysql -uroot -p123456
# 切换目录
cd /etc/mysql
# 安装 vim
apt-get update
apt-get install vim
# 修改配置
vim my.cnf
```

步骤 02 新增下面的内容：

```
[mysqld]
## 同一个局域网内注意要唯一
server-id=100
## 开启二进制日志功能
log-bin=master-bin
binlog-format=ROW           # 二进制日志格式有 3 种，即 row、statement、mixed
# binlog-do-db=数据库名       # 同步的数据库名称，如果不配置，则表示同步所有的库
```

步骤 03 重新启动：

```
service mysql restart
docker restart mysql-master
```

步骤 04 创建数据库同步账户，其用户名为从节点，密码为 123456：

```
docker exec -it mysql-master /bin/bash
mysql -uroot -p123456
create user 'slave'@'%' identified by '123456';
grant replicaiton slave, replication client on *.* to 'slave'@'%';
```

17.3.2　配置从节点

配置从节点的操作流程如下：

步骤 01 执行如下命令进入从节点容器内部：

```
# 进入从节点容器
docker exec -it mysql-slave /bin/bash
# 登录 MySQL
mysql -uroot -p123456
# 切换目录
cd /etc/mysql
# 安装 vim
apt-get update
apt-get install vim
# 修改配置
vim my.cnf
```

步骤 02 新增下面的内容：

```
[mysqld]
## 设置 server_id，注意要唯一
server-id=101
```

```
## 开启二进制日志功能，以备从节点作为其他从节点的主节点时使用
log-bin=mysql-slave-bin
## relay_log 配置中继日志
relay_log=mysql-relay-bin
read_only=1    ## 设置为只读，该项如果不设置，则表示从节点可读可写
```

步骤03 重新启动数据库服务：

```
service mysql restart
docker restart mysql-slave
```

17.3.3 主从复制

进行主从复制的操作流程如下：

步骤01 进入主节点：

```
# 进入主节点
docker exec -it mysql-master /bin/bash
mysql -uroot -p123456
```

步骤02 进入主节点库的 MySQL 客户端，输入 show master status 查看主节点的状态，其结构如图 17-3 所示。

图 17-3 主节点的日志偏移量的信息查询

需要记住当前 File 的值和 Position 的值，设置主从同步时需要使用。

步骤03 查询主节点的 IP 地址。

进入从节点库的 MySQL 客户端，执行如下命令：

```
# 进入从节点
docker exec -it mysql-slave /bin/bash
# 登录 MySQL
mysql -uroot -p123456
# 设置同步，需要注意连接主节点的 IP 地址和端口号，以及日志文件名称和日志偏移量
change master to master_host='47.95.2.2', master_user='slave', master_password='123456', master_port=3301, master_log_file='master-bin.000001', master_log_pos=617, master_connect_retry=30;
```

步骤04 使用 "start slave;" 开启主从复制过程，然后再次执行 show slave status \G 语句查询主从同步状态，结果如图 17-4 所示。

```
# 验证
start slave;
show slave status \G
```

图 17-4 主从架构的验证

Slave_IO_Running 和 Slave_SQL_Running 是查看主从服务器是否运行的关键字段，默认为 NO，表示没有进行主从复制。如果 Slave_IO_Running 和 Slave_SQL_Running 的值都是 Yes，则说明搭建成功。

17.4 本章练习

1. 举例说明读写分离架构的应用场景。
2. 举例说明读写架构对于数据即时查询的业务有何影响。

第 8 部分　MySQL 操作实战

第 18 章　MySQL 数据操作实战

第 18 章

MySQL 数据操作实战

本章内容：

- 数据库和数据表的创建与查看
- 新增、修改、删除表数据
- MySQL 基础查询和 where 子查询
- MySQL Like 模糊查询
- MySQL 分组查询、聚合函数、排序查询
- MySQL 关联查询
- MySQL 子查询、多表查询

本章将讲解 MySQL 数据操作实战，内容包括数据库和数据表的创建与查询，新增、修改、删除表数据，基础查询和 where 子查询，like 模糊查询，分组查询，聚合函数，排序查询，关联查询，子查询，多表查询，等等，并通过实例一一向读者进行展示。

18.1　数据库和数据表的创建与查看

本节演示数据库、数据表的创建与查看的实际操作。

例 18.1　查看 MySQL 服务器中所有的数据库，示例 SQL 语句及其执行结果如下：

```
mysql> show databases;
+--------------------+
| Database           |
+--------------------+
```

```
| information_schema |
| abc                |
| mysql              |
| performance_schema |
| sys                |
+--------------------+
5 rows in set (0.00 sec)
```

从结果中可以看出，当前 MySQL 服务器中有 5 个数据库。

例 18.2　进入某一数据库（只有进入数据库后，才能操作库中的表和表记录），示例 SQL 语句及其执行结果如下：

```
# 进入名为 test 的数据库
mysql> use test;
Database changed
mysql> --
# 查询当前操作的数据库名称
mysql> select database();
+------------+
| database() |
+------------+
| test       |
+------------+
1 row in set (0.00 sec)
```

从结果中可以看出，已经进入 test 数据库了。

例 18.3　查看 mysql 数据库中的所有表，示例 SQL 语句及其执行结果如下：

```
mysql> show tables;
+---------------------------+
| Tables_in_mysql           |
+---------------------------+
| columns_priv              |
| db                        |
| engine_cost               |
| event                     |
| func                      |
| general_log               |
| gtid_executed             |
| help_category             |
| help_keyword              |
| help_relation             |
| help_topic                |
| innodb_index_stats        |
| innodb_table_stats        |
| ndb_binlog_index          |
| plugin                    |
| proc                      |
| procs_priv                |
| proxies_priv              |
| server_cost               |
| servers                   |
```

```
| slave_master_info           |
| slave_relay_log_info        |
| slave_worker_info           |
| slow_log                    |
| tables_priv                 |
| time_zone                   |
| time_zone_leap_second       |
| time_zone_name              |
| time_zone_transition        |
| time_zone_transition_type   |
| user                        |
+-----------------------------+
31 rows in set (0.00 sec)
```

从结果中可以看出，当前 mysql 数据库中有 31 张表。

例 18.4 删除 mydb 数据库名，示例 SQL 语句及其执行结果如下：

```
# 直接删除，因为不存在此名称的数据库，所以报错
mysql> drop database mydb;
ERROR 1008 (HY000): Can't drop database 'mydb'; database doesn't exist
mysql> /* */
# 使用此种语法可以避免删除不存在的数据库导致的错误
mysql> drop database if exists mydb;
Query OK, 0 rows affected, 1 warning (0.00 sec)

mysql> -- mydb
```

例 18.5 重新创建 mydb 数据库，指定编码为 utf8，示例 SQL 语句及其执行结果如下：

```
# 创建一个名为"mydb"的新数据库，并将该数据库的默认字符集设置为 UTF-8 编码
mysql> create database mydb charset utf8;
Query OK, 1 row affected (0.00 sec)
# 指定当前数据库的默认字符集为 ISO-8859-1 编码。这个命令不会对已经存在的数据库产生影响
mysql>
mysql> -- latin1/iso8859-1,
# 退出当前 MySQL 客户端的状态，以便回到上一个提示符下继续操作
mysql>
mysql> -- mydb;
```

例 18.6 查看建库时的语句（并验证数据库使用的编码），示例 SQL 语句及其执行结果如下：

```
mysql> show create database mydb;
+----------+-----------------------------------------------------------------+
| Database | Create Database                                                 |
+----------+-----------------------------------------------------------------+
| mydb     | CREATE DATABASE `mydb` /*!40100 DEFAULT CHARACTER SET utf8 */   |
+----------+-----------------------------------------------------------------+
1 row in set (0.00 sec)
```

这段代码是用于在 MySQL 中查看创建数据库的语句。具体来说，show create database mydb; 命令会输出创建名为"mydb"的数据库的完整语句。执行结果中，第一列是数据库名；第二列

是 SQL 语句，CREATE DATABASE 表示创建数据库的关键字，mydb 则是要创建的数据库名，接下来的部分是 SQL 语句的具体内容，包括字符集、默认值等选项，在这个例子中，字符集被设置为 UTF-8 编码，这是 MySQL 中常用的一种字符集。

例 18.7 进入 mydb 库，删除 student 表（如果存在），示例 SQL 语句如下：

```
mysql> use mydb;
Database changed
mysql> drop table if exists student;
Query OK, 0 rows affected, 1 warning (0.00 sec)
```

例 18.8 创建 student 表（编号[数值类型]、姓名、性别、出生年月、考试成绩[浮点型]），示例 SQL 语句如下：

```
drop table if exists student;
    create table student(
        id int primary key auto_increment,
        name varchar(50),
        gender varchar(2),
        birthday date,
        score double
    );
```

例 18.9 查看 student 表结构，示例 SQL 语句及其执行结果如下：

```
mysql> desc student;
+----------+-------------+------+-----+---------+----------------+
| Field    | Type        | Null | Key | Default | Extra          |
+----------+-------------+------+-----+---------+----------------+
| id       | int(11)     | NO   | PRI | NULL    | auto_increment |
| name     | varchar(50) | YES  |     | NULL    |                |
| gender   | varchar(2)  | YES  |     | NULL    |                |
| birthday | date        | YES  |     | NULL    |                |
| score    | double      | YES  |     | NULL    |                |
+----------+-------------+------+-----+---------+----------------+
5 rows in set (0.00 sec)
```

由结果可以看出，创建的 student 表符合预期。

例 18.10 查看建表时的语句，示例 SQL 语句及其执行结果如下：

```
mysql> show create tables student;
| student | CREATE TABLE `student` (
  `id` int(11) NOT NULL AUTO_INCREMENT,
  `name` varchar(50) DEFAULT NULL,
  `gender` varchar(20) DEFAULT NULL,
  `birthday` date DEFAULT NULL,
  `score` double DEFAULT NULL,
  PRIMARY KEY (`id`)
) ENGINE=InnoDB DEFAULT CHARSET=utf8 |
1 row in set (0.00 sec)
```

这段代码是用于在 MySQL 中查看创建名为"student"的表的完整语句。具体来说，show

create tables student;命令会输出创建名为"student"的表的完整语句。执行结果中，第一列是表名；第二列是 SQL 语句，CREATE TABLE 表示创建表的关键字，student 则是要创建的表名，接下来的部分是 SQL 语句的具体内容，包括各个字段的名称、数据类型、默认值和是否为主键等选项。在这个例子中，student 表包含了 5 个字段：id、name、gender、birthday 和 score，分别使用了不同的数据类型进行存储。其中，id 字段被指定为主键，AUTO_INCREMENT 属性表示该字段会自动递增。

18.2 新增、修改、删除表数据

本节演示表数据的新增、修改、删除操作。

例 18.11 往 student 表中插入记录，示例 SQL 语句及其执行结果如下：

```
mysql> insert into student(name,gender,birthday,score) values('zhangsan','m','1999-2-2',70);
Query OK, 1 row affected (0.00 sec)

mysql> insert into student values(null,'lisi','m','1997-2-2',80);
Query OK, 1 row affected (0.00 sec)

mysql> insert into student values(null,'wangwu','m','1989-2-2',75);
Query OK, 1 row affected (0.00 sec)
```

在 student 表中插入了 3 条记录。

例 18.12 查询 student 表中所有学生的信息，示例 SQL 语句及其执行结果如下：

```
mysql> select * from student;
+----+----------+--------+------------+-------+
| id | name     | gender | birthday   | score |
+----+----------+--------+------------+-------+
|  1 | zhangsan | m      | 1999-02-02 |    70 |
|  2 | lisi     | m      | 1997-02-02 |    80 |
|  3 | wangwu   | m      | 1989-02-02 |    75 |
+----+----------+--------+------------+-------+
3 rows in set (0.00 sec)
```

例 18.13 修改 student 表中所有学生的成绩，加 10 分特长分，示例 SQL 语句及其执行结果如下：

```
mysql>  update student set score=score+10;
Query OK, 3 rows affected (0.01 sec)
Rows matched: 3  Changed: 3  Warnings: 0
```

MySQL 返回一个"Query OK"的状态码，表示操作成功。接着显示受影响的行数、匹配的行数和修改的行数等信息。在本例中，受影响的行数为 3，说明有 3 行数据被更新了；匹配的行数也是 3，说明更新的行数正好是查询结果中的 3 行；修改的行数也是 3，说明这 3 行数据都被成功更新了。最后，没有输出任何警告信息。

例 18.14 修改 student 表中 zhangsan 的成绩，将成绩改为 98 分，示例 SQL 语句及其执行结果如下：

```
mysql>    update student set score=98 where name='zhangsan';
Query OK, 1 row affected (0.00 sec)
Rows matched: 1  Changed: 1  Warnings: 0
```

MySQL 返回一个"Query OK"的状态码，表示操作成功，将 student 表中所有 name 字段为 zhangsan 的行的 score 字段值改为 98。

例 18.15 删除表数据，示例 SQL 语句及其执行结果如下：

```
# 删除性别是 w 的数据
mysql> delete from student  where gender='w';
Query OK, 0 rows affected (0.00 sec)
# 删除表中所有数据，数据还能找回
mysql>    delete from student;
Query OK, 0 rows affected (0.00 sec)
# 清空表数据，效率高，但是数据找不回
mysql>    truncate student ;
Query OK, 0 rows affected (0.00 sec)
```

18.3 基础查询和 where 子句查询

本节演示 MySQL 基础查询和 Where 子句查询的实际操作。

首先准备数据。

例 18.16 准备部门表数据和员工表数据，示例 SQL 语句如下：

```
# 创建部门表，字段依次为部门编号、部门名称、部门地址
drop table if exists dept ;
 create table dept (
 deptno int(2) not null,
dname varchar(14) collate utf8_bin default null,
   loc varchar(13) collate utf8_bin default null,
   primary key (deptno)
   ) engine=innodb default charset=utf8 collate=utf8_bin;
Query OK, 0 rows affected (0.00 sec)

# 插入表数据
 insert into dept values(10, 'accounting', 'new york');
 insert into dept values(20, 'research','dallas');
 insert into dept values(30,'sales', 'chicago');
 insert into dept values(40, 'operations','boston');
# 创建员工表，字段依次为员工号、员工性别、出生日期、入职日期、薪资、奖金、职位、部门编号
drop table if exists emp;
```

```sql
create table emp
    (
    eno integer not null ,
    ename varchar(20) not null,
    sex varchar(20) not null ,
    birthday varchar(20) not null ,
    jdate varchar(20) not null ,
    salary integer not null ,
bonus integer not null ,
    epost varchar(20) not null,
    deptno int(2)not null,
    primary key (eno)
    );

insert into emp(eno , ename , sex , birthday , jdate , salary ,bonus, epost ,deptno )values
    (2,'wanger','m','1985-5-20','1993-2-11','8000','500','worker',10);
insert into emp(eno , ename , sex , birthday , jdate , salary ,bonus, epost ,deptno )values
    (3,'libai','w','1978-7-4','1999-10-18','4500','500','worker',10);
insert into emp(eno , ename , sex , birthday , jdate , salary ,bonus, epost ,deptno )values
    (4,'lisi','w','1987-3-15','2006-7-10','6500','500','worker',20);
insert into emp(eno , ename , sex , birthday , jdate , salary ,bonus, epost ,deptno )values
    (5,'litaibai','m','1979-1-19','2010-9-13','6000','500','worker',30);
insert into emp(eno , ename , sex , birthday , jdate , salary ,bonus, epost ,deptno )values
    (6,'wangming','m','1978-9-29','2012-5-1','7500','1500','manager',30);
insert into emp(eno , ename , sex , birthday , jdate , salary ,bonus, epost ,deptno )values
    (7,'zhoulei','w','1989-12-6','1999-3-23','5000','500','worker',40);
insert into emp(eno , ename , sex , birthday , jdate , salary ,bonus, epost ,deptno )values
    (8,'xiaoli','m','1990-11-11','20013-2-12','4800','500','worker',40);
```

接下来开始查询。

例 18.17 查询 emp 表中的所有员工,显示姓名、薪资、奖金,示例 SQL 语句及其执行结果如下:

```
mysql> select ename, salary , bonus
    ->    from emp;
+----------+--------+-------+
| ename    | salary | bonus |
+----------+--------+-------+
| wanger   |   8000 |   500 |
| libai    |   4500 |   500 |
| lisi     |   6500 |   500 |
| litaibai |   6000 |   500 |
| wangming |   7500 |  1500 |
| zhoulei  |   5000 |   500 |
| xiaoli   |   4800 |   500 |
+----------+--------+-------+
```

```
7 rows in set (0.00 sec)
```

结果中显示了 emp 表中所有员工的姓名、薪资和奖金数据。

例 18.18 查询 emp 表中的所有部门和职位,示例 SQL 语句及其执行结果如下:

```
mysql> select deptno , epost from emp;
+--------+---------+
| deptno | epost   |
+--------+---------+
|     10 | worker  |
|     10 | worker  |
|     20 | worker  |
|     30 | worker  |
|     30 | manager |
|     40 | worker  |
|     40 | worker  |
+--------+---------+
7 rows in set (0.00 sec)
```

结果中显示了 emp 表中的所有部门和职位数据。

例 18.19 查询 emp 表中的所有部门和职位,并对数据去重,示例 SQL 语句及其执行结果如下:

```
mysql> select distinct deptno , epost from emp;
+--------+---------+
| deptno | epost   |
+--------+---------+
|     10 | worker  |
|     20 | worker  |
|     30 | worker  |
|     30 | manager |
|     40 | worker  |
+--------+---------+
5 rows in set (0.00 sec)
```

结果中显示了 emp 表中不重复的部门和职位数据。

例 18.20 查询 emp 表中薪资大于 5000 的所有员工,显示员工姓名、薪资,示例 SQL 语句及其执行结果如下:

```
mysql> select ename,salary from emp
    -> where salary >5000;
+----------+--------+
| ename    | salary |
+----------+--------+
| wanger   |   8000 |
| lisi     |   6500 |
| litaibai |   6000 |
| wangming |   7500 |
+----------+--------+
4 rows in set (0.00 sec)
```

结果中显示了 emp 表中薪资大于 5000 的所有员工的姓名、薪资数据。

例 18.21 查询 emp 表中总薪资（薪资+奖金）大于 6500 的所有员工，显示员工姓名、总薪资，示例 SQL 语句及其执行结果如下：

```
mysql> select ename, salary +bonus from emp where salary +bonus > 6500;
+----------+---------------+
| ename    | salary +bonus |
+----------+---------------+
| wanger   |          8500 |
| lisi     |          7000 |
| wangming |          9000 |
+----------+---------------+
3 rows in set (0.00 sec)
```

结果中显示了 emp 表中总薪资大于 6500 的所有员工的姓名、总薪资数据。

例 18.22 注意例 18.20 的查询结果中的表头，将表头中的"salavy+bonus"修改为"total-salary"，示例 SQL 语句及其执行结果如下：

```
mysql> select ename, salary +bonus 'total-salary' from emp where salary +bonus > 6500;
+----------+--------------+
| ename    | total-salary |
+----------+--------------+
| wanger   |         8500 |
| lisi     |         7000 |
| wangming |         9000 |
+----------+--------------+
3 rows in set (0.00 sec)
```

结果中显示成功将例 18.20 的表头中的"salary+bonus"修改为"total-salary"。

例 18.23 查询 emp 表中薪资在 7000 和 10000 之间的员工，显示员工姓名和薪资，示例 SQL 语句及其执行结果如下：

```
# 普通写法
mysql> select ename, salary from emp where salary >=7000 and salary <=10000;
+----------+--------+
| ename    | salary |
+----------+--------+
| wanger   |   8000 |
| wangming |   7500 |
+----------+--------+
2 rows in set (0.00 sec)

# 使用 between and 的写法
mysql> select ename, salary from emp where salary between 7000 and 10000;
+----------+--------+
| ename    | salary |
+----------+--------+
| wanger   |   8000 |
| wangming |   7500 |
```

```
+----------+--------+
2 rows in set (0.00 sec)
```

上述两种写法都成功查询出了 emp 表中薪资在 7000 和 10000 之间的员工。

例 18.24 查询 emp 表中薪资为 5000、6000、8000 的员工，显示员工姓名和薪资，示例 SQL 语句及其执行结果如下：

```
# 普通写法
mysql> select ename, salary from emp where salary =5000 or salary =6000 or salary =8000;
+----------+--------+
| ename    | salary |
+----------+--------+
| wanger   |   8000 |
| litaibai |   6000 |
| zhoulei  |   5000 |
+----------+--------+
3 rows in set (0.00 sec)

# 使用 in 的写法
mysql> select ename, salary from emp where salary in(5000,6000,8000);
+----------+--------+
| ename    | salary |
+----------+--------+
| wanger   |   8000 |
| litaibai |   6000 |
| zhoulei  |   5000 |
+----------+--------+
3 rows in set (0.00 sec)
```

上述两种写法都成功查询出了 emp 表中薪资为 5000、6000、8000 的员工。

例 18.25 查询 emp 表中薪资不为 5000、6000、8000 的员工，显示员工姓名和薪资，示例 SQL 语句及其执行结果如下：

```
# 普通写法
mysql> select ename, salary from emp where not( salary =5000 or salary =6000 or salary =8000);
+----------+--------+
| ename    | salary |
+----------+--------+
| libai    |   4500 |
| lisi     |   6500 |
| wangming |   7500 |
| xiaoli   |   4800 |
+----------+--------+
4 rows in set (0.00 sec)
# 使用 not in 的写法
mysql> select ename, salary from emp where salary not in(5000,6000,8000);
+----------+--------+
| ename    | salary |
+----------+--------+
| libai    |   4500 |
```

```
| lisi     | 6500 |
| wangming | 7500 |
| xiaoli   | 4800 |
+----------+------+
4 rows in set (0.01 sec)
```

上述两种写法都成功查询出了 emp 表中薪资不为 5000、6000、8000 的员工。

例 18.26 查询 emp 表中薪资大于 4000 和薪资小于 2000 的员工，显示员工姓名、薪资，示例 SQL 语句及其执行结果如下：

```
mysql> select ename, salary sal from emp where salary >4000 or salary <2000;
+----------+------+
| ename    | sal  |
+----------+------+
| wanger   | 8000 |
| libai    | 4500 |
| lisi     | 6500 |
| litaibai | 6000 |
| wangming | 7500 |
| zhoulei  | 5000 |
| xiaoli   | 4800 |
+----------+------+
7 rows in set (0.00 sec)
```

结果显示成功查询出了 emp 表中薪资大于 4000 和薪资小于 2000 的员工。

18.4 like 模糊查询

本节演示 like 模糊查询的实际操作。

例 18.27 查询 emp 表中姓名中以 "li" 开头的员工，显示员工姓名，示例 SQL 语句及其执行结果如下：

```
mysql> select ename, salary sal from emp where ename like 'li%';
+----------+------+
| ename    | sal  |
+----------+------+
| libai    | 4500 |
| lisi     | 6500 |
| litaibai | 6000 |
+----------+------+
3 rows in set (0.00 sec)
```

结果中显示出了 emp 表中姓名以 "li" 开头的员工的姓名。

例 18.28 查询 emp 表中姓名中包含 "li" 的员工，显示员工姓名，示例 SQL 语句及其执行结果如下：

```
mysql> select ename, salary sal from emp where ename like '%li%';
+----------+------+
| ename    | sal  |
+----------+------+
| libai    | 4500 |
| lisi     | 6500 |
| litaibai | 6000 |
| xiaoli   | 4800 |
+----------+------+
4 rows in set (0.00 sec)
```

结果中显示出了 emp 表中姓名中包含 "li" 的员工的姓名。

例 18.29　查询 emp 表中姓名以 "li" 结尾的员工，显示员工姓名，示例 SQL 语句及其执行结果如下：

```
mysql> select ename, salary sal from emp where ename like '%li';
+--------+------+
| ename  | sal  |
+--------+------+
| xiaoli | 4800 |
+--------+------+
1 row in set (0.00 sec)
```

结果中显示出了 emp 表中姓名以 "li" 结尾的员工的姓名。

18.5　分组查询、聚合函数、排序查询

本节演示分组查询、聚合函数、排序查询的实际操作。

例 18.30　对 emp 表按照职位进行分组，并统计每个职位的人数，显示职位和对应人数，示例 SQL 语句及其执行结果如下：

```
mysql> select epost ,count(*) from emp  group by epost ;
+---------+----------+
| epost   | count(*) |
+---------+----------+
| manager |        1 |
| worker  |        6 |
+---------+----------+
2 rows in set (0.00 sec)
```

结果符合预期。

例 18.31　按照部门分组，显示部门、最高薪资，示例 SQL 语句及其执行结果如下：

```
mysql> select deptno, max(salary ) from emp group by deptno;
+--------+--------------+
| deptno | max(salary ) |
+--------+--------------+
```

```
|   10   |    8000    |
|   20   |    6500    |
|   30   |    7500    |
|   40   |    5000    |
+--------+------------+
4 rows in set (0.00 sec)
```

结果符合预期。

例 18.32 查询每个部门的最高薪资、显示部门、员工姓名、最高薪资，示例 SQL 语句及其执行结果如下：

```
mysql> select emp.deptno , ename, t1.msal from emp,(select deptno ,max(salary ) msal
from emp group by deptno ) t1   where emp.deptno = t1.deptno and emp.salary =t1.msal;
+--------+----------+------+
| deptno | ename    | msal |
+--------+----------+------+
|   10   | wanger   | 8000 |
|   20   | lisi     | 6500 |
|   30   | wangming | 7500 |
|   40   | zhoulei  | 5000 |
+--------+----------+------+
4 rows in set (0.00 sec)
```

结果符合预期。

例 18.33 统计 emp 表中薪资大于 3000 的员工个数，示例 SQL 语句及其执行结果如下：

```
mysql> select count(eno) from emp where salary >3000;
+------------+
| count(eno) |
+------------+
|      7     |
+------------+
1 row in set (0.00 sec)
```

结果符合预期。

例 18.34 统计 emp 表中所有员工的薪资总和（不包含奖金），示例 SQL 语句及其执行结果如下：

```
mysql> select sum(salary ) from emp;
+--------------+
| sum(salary ) |
+--------------+
|    42300     |
+--------------+
1 row in set (0.01 sec)
```

结果符合预期。

例 18.35 统计 emp 表中员工的平均薪资(不包含奖金)，示例 SQL 语句及其执行结果如下：

```
# 普通方式计算平均数
```

```
mysql> select sum(salary )/count(*) from emp;
+---------------------+
| sum(salary )/count(*) |
+---------------------+
|              6042.8571 |
+---------------------+
1 row in set (0.00 sec)
# 使用 avg 函数求平均数
mysql> select avg(salary ) from emp;
+--------------+
| avg(salary ) |
+--------------+
|    6042.8571 |
+--------------+
1 row in set (0.00 sec)
```

上述两种计算方式等效。

例 18.36 查询 emp 表中所有在 1978 年和 1985 年之间出生的员工，显示姓名、出生日期，示例 SQL 语句及其执行结果如下：

```
mysql> select ename, birthday from emp where year(birthday) between 1978 and 1985;
+----------+-----------+
| ename    | birthday  |
+----------+-----------+
| wanger   | 1985-5-20 |
| libai    | 1978-7-4  |
| litaibai | 1979-1-19 |
| wangming | 1978-9-29 |
+----------+-----------+
4 rows in set (0.00 sec)
```

结果显示查询出了所有在 1978 年和 1985 年之间出生的员工。

例 18.37 查询要在本月过生日的所有员工，示例 SQL 语句及其执行结果如下：

```
mysql> select * from emp where month(curdate()) = month(birthday);
Empty set (0.00 sec)
```

结果显示本月没有员工过生日。

例 18.38 对 emp 表中所有员工的薪资进行升序（从低到高）排序，显示员工姓名、薪资，示例 SQL 语句及其执行结果如下：

```
# 默认就是升序排序，所以 asc 可以省略不写
mysql>  select ename, salary from emp order by salary ;
+----------+--------+
| ename    | salary |
+----------+--------+
| libai    |   4500 |
| xiaoli   |   4800 |
| zhoulei  |   5000 |
| litaibai |   6000 |
```

```
| lisi     | 6500 |
| wangming | 7500 |
| wanger   | 8000 |
+----------+------+
7 rows in set (0.00 sec)
```

结果符合预期。

例 18.39 对 emp 表中所有员工奖金进行降序（从高到低）排序，显示员工姓名、奖金，示例 SQL 语句及其执行结果如下：

```
mysql> select ename, bonus from emp order by bonus desc;
+----------+-------+
| ename    | bonus |
+----------+-------+
| wangming | 1500  |
| wanger   | 500   |
| libai    | 500   |
| lisi     | 500   |
| litaibai | 500   |
| zhoulei  | 500   |
| xiaoli   | 500   |
+----------+-------+
7 rows in set (0.00 sec)
```

结果符合预期。

例 18.40 查询 emp 表中的所有记录，分页显示每页显示（3 条记录），返回第 1 页，示例 SQL 语句及其执行结果如下：

```
# limit 使用语法：limit (需要查询的页码数-1)*每页需要返回的行数，返回的具体行数
mysql> select * from emp limit 0,3;
+-----+--------+-----+-----------+------------+--------+-------+--------+--------+
| eno | ename  | sex | birthday  | jdate      | salary | bonus | epost  | deptno |
+-----+--------+-----+-----------+------------+--------+-------+--------+--------+
| 2   | wanger | m   | 1985-5-20 | 1993-2-11  | 8000   | 500   | worker | 10     |
| 3   | libai  | w   | 1978-7-4  | 1999-10-18 | 4500   | 500   | worker | 10     |
| 4   | lisi   | w   | 1987-3-15 | 2006-7-10  | 6500   | 500   | worker | 20     |
+-----+--------+-----+-----------+------------+--------+-------+--------+--------+
3 rows in set (0.00 sec)
```

limit 子句用于限制从 select 语句返回的行数。由于是返回第 1 页，每页 3 行数据，因此设置为 limit 0,3，因为 limit 0 表示从记录行偏移量为 0 的数据行开始查询。如果不指定 limit 子句，则默认从第一条记录开始，并显示所有记录。

例 18.41 查询 emp 表中的所有记录，分页显示（每页显示 3 条记录），返回第 2 页，示例 SQL 语句及其执行结果如下：

```
# limit (页码-1)*每页显示条数, 每页显示条数
mysql> select * from emp limit 3,3;
```

```
+-----+----------+-----+-----------+-----------+--------+------+--------+------+
| eno | ename    | sex | birthday  | jdate     | salary | bonus| epost  |deptno|
+-----+----------+-----+-----------+-----------+--------+------+--------+------+
|   5 | litaibai | m   | 1979-1-19 | 2010-9-13 |   6000 |  500 | worker |  30  |
|   6 | wangming | m   | 1978-9-29 | 2012-5-1  |   7500 | 1500 | manager|  30  |
|   7 | zhoulei  | w   | 1989-12-6 | 1999-3-23 |   5000 |  500 | worker |  40  |
+-----+----------+-----+-----------+-----------+--------+------+--------+------+
3 rows in set (0.00 sec)
```

18.6 关联查询

本节演示 MySQL 关联查询的实际操作：

例 18.42 查询部门和部门对应的员工信息，示例 SQL 语句及其执行结果如下：

```
# 笛卡儿积查询：如果有两张表，其中一张表有 m 条数据，另外一张表有 n 条数据，则笛卡儿积查询的结果是
m*n 条，这种查询中包含大量错误的数据，一般我们不会直接使用这种查询
mysql> select * from dept,emp;
+--------+------------+----------+-----+----------+-----+-----------+------------+--------+------+--------+--------+
| deptno | dname      | loc      | eno | ename    | sex | birthday  | jdate      | salary | bonus| epost  | deptno |
+--------+------------+----------+-----+----------+-----+-----------+------------+--------+------+--------+--------+
|   10   | accounting | new york |  2  | wanger   | m   | 1985-5-20 | 1993-2-11  |  8000  |  500 | worker |   10   |
|   20   | research   | dallas   |  2  | wanger   | m   | 1985-5-20 | 1993-2-11  |  8000  |  500 | worker |   10   |
|   30   | sales      | Chicago  |  2  | wanger   | m   | 1985-5-20 | 1993-2-11  |  8000  |  500 | worker |   10   |
|   40   | operations | boston   |  2  | wanger   | m   | 1985-5-20 | 1993-2-11  |  8000  |  500 | worker |   10   |
|   10   | accounting | new york |  3  | libai    | w   | 1978-7-4  | 1999-10-18 |  4500  |  500 | worker |   10   |
|   20   | research   | dallas   |  3  | libai    | w   | 1978-7-4  | 1999-10-18 |  4500  |  500 | worker |   10   |
|   30   | sales      | Chicago  |  3  | libai    | w   | 1978-7-4  | 1999-10-18 |  4500  |  500 | worker |   10   |
|   40   | operations | boston   |  3  | libai    | w   | 1978-7-4  | 1999-10-18 |  4500  |  500 | worker |   10   |
|   10   | accounting | new york |  4  | lisi     | w   | 1987-3-15 | 2006-7-10  |  6500  |  500 | worker |   20   |
|   20   | research   | dallas   |  4  | lisi     | w   | 1987-3-15 | 2006-7-10  |  6500  |  500 | worker |   20   |
|   30   | sales      | Chicago  |  4  | lisi     | w   | 1987-3-15 | 2006-7-10  |  6500  |  500 | worker |   20   |
|   40   | operations | boston   |  4  | lisi     | w   | 1987-3-15 | 2006-7-10  |  6500  |  500 | worker |   20   |
|   10   | accounting | new york |  5  | litaibai | m   | 1979-1-19 | 2010-9-13  |  6000  |  500 | worker |   30   |
|   20   | research   | dallas   |  5  | litaibai | m   | 1979-1-19 | 2010-9-13  |  6000  |  500 | worker |   30   |
|   30   | sales      | Chicago  |  5  | litaibai | m   | 1979-1-19 | 2010-9-13  |  6000  |  500 | worker |   30   |
|   40   | operations | boston   |  5  | litaibai | m   | 1979-1-19 | 2010-9-13  |  6000  |  500 | worker |   30   |
|   10   | accounting | new york |  6  | wangming | m   | 1978-9-29 | 2012-5-1   |  7500  | 1500 | manager|   30   |
|   20   | research   | dallas   |  6  | wangming | m   | 1978-9-29 | 2012-5-1   |  7500  | 1500 | manager|   30   |
|   30   | sales      | Chicago  |  6  | wangming | m   | 1978-9-29 | 2012-5-1   |  7500  | 1500 | manager|   30   |
|   40   | operations | boston   |  6  | wangming | m   | 1978-9-29 | 2012-5-1   |  7500  | 1500 | manager|   30   |
|   10   | accounting | new york |  7  | zhoulei  | w   | 1989-12-6 | 1999-3-23  |  5000  |  500 | worker |   40   |
|   20   | research   | dallas   |  7  | zhoulei  | w   | 1989-12-6 | 1999-3-23  |  5000  |  500 | worker |   40   |
|   30   | sales      | chicago  |  7  | zhoulei  | w   | 1989-12-6 | 1999-3-23  |  5000  |  500 | worker |   40   |
|   40   | operations | boston   |  7  | zhoulei  | w   | 1989-12-6 | 1999-3-23  |  5000  |  500 | worker |   40   |
|   10   | accounting | new york |  8  | xiaoli   | m   | 1990-11-11| 20013-2-12 |  4800  |  500 | worker |   40   |
|   20   | research   | dallas   |  8  | xiaoli   | m   | 1990-11-11| 20013-2-12 |  4800  |  500 | worker |   40   |
|   30   | sales      | chicago  |  8  | xiaoli   | m   | 1990-11-11| 20013-2-12 |  4800  |  500 | worker |   40   |
|   40   | operations | boston   |  8  | xiaoli   | m   | 1990-11-11| 20013-2-12 |  4800  |  500 | worker |   40   |
+--------+------------+----------+-----+----------+-----+-----------+------------+--------+------+--------+--------+
28 rows in set (0.00 sec)
```

```
# 通过关联条件，提取大量错误的数据，保留正确的数据，这就是关联查询，其中 where 后面的条件就是关联条件
mysql> select * from dept,emp where dept.deptno = emp.deptno ;
+--------+------------+----------+-----+----------+-----+-----------+------------+--------+-------+---------+--------+
| deptno | dname      | loc      | eno | ename    | sex | birthday  | jdate      | salary | bonus | epost   | deptno |
+--------+------------+----------+-----+----------+-----+-----------+------------+--------+-------+---------+--------+
| 10     | accounting | new york | 2   | wanger   | m   | 1985-5-20 | 1993-2-11  | 8000   | 500   | worker  | 10     |
| 10     | accounting | new york | 3   | libai    | w   | 1978-7-4  | 1999-10-18 | 4500   | 500   | worker  | 10     |
| 20     | research   | dallas   | 4   | lisi     | w   | 1987-3-15 | 2006-7-10  | 6500   | 500   | worker  | 20     |
| 30     | sales      | chicago  | 5   | litaibai | m   | 1979-1-19 | 2010-9-13  | 6000   | 500   | worker  | 30     |
| 30     | sales      | chicago  | 6   | wangming | m   | 1978-9-29 | 2012-5-1   | 7500   | 1500  | manager | 30     |
| 40     | operations | boston   | 7   | zhoulei  | w   | 1989-12-6 | 1999-3-23  | 5000   | 500   | worker  | 40     |
| 40     | operations | boston   | 8   | xiaoli   | m   | 1990-11-11| 20013-2-12 | 4800   | 500   | worker  | 40     |
+--------+------------+----------+-----+----------+-----+-----------+------------+--------+-------+---------+--------+
7 rows in set (0.00 sec)
```

通过关联查询，可以轻松地从多张表中提取所需的数据，并将它们合并到一个查询结果中。

例 18.43 查询所有部门和部门下的员工，如果部门下没有员工，则员工显示为 null（一定要列出所有部门），示例 SQL 语句及其执行结果如下：

```
# 插入一条部门的测试信息
mysql> insert into dept values(50, 'accounting2', 'new york2');
Query OK, 1 row affected (0.00 sec)
# left join:左外连接查询，会将左边表中的所有数据都列出来，右边表只显示符合条件的记录
mysql> select * from dept left join emp on dept.deptno = emp.deptno;
+--------+-------------+-----------+------+----------+------+-----------+------------+--------+-------+---------+--------+
| deptno | dname       | loc       | eno  | ename    | sex  | birthday  | jdate      | salary | bonus | epost   | deptno |
+--------+-------------+-----------+------+----------+------+-----------+------------+--------+-------+---------+--------+
| 10     | accounting  | new york  | 2    | wanger   | m    | 1985-5-20 | 1993-2-11  | 8000   | 500   | worker  | 10     |
| 10     | accounting  | new york  | 3    | libai    | w    | 1978-7-4  | 1999-10-18 | 4500   | 500   | worker  | 10     |
| 20     | research    | dallas    | 4    | lisi     | w    | 1987-3-15 | 2006-7-10  | 6500   | 500   | worker  | 20     |
| 30     | sales       | chicago   | 5    | litaibai | m    | 1979-1-19 | 2010-9-13  | 6000   | 500   | worker  | 30     |
| 30     | sales       | chicago   | 6    | wangming | m    | 1978-9-29 | 2012-5-1   | 7500   | 1500  | manager | 30     |
| 40     | operations  | boston    | 7    | zhoulei  | w    | 1989-12-6 | 1999-3-23  | 5000   | 500   | worker  | 40     |
| 40     | operations  | boston    | 8    | xiaoli   | m    | 1990-11-11| 20013-2-12 | 4800   | 500   | worker  | 40     |
| 50     | accounting2 | new york2 | NULL | NULL     | NULL | NULL      | NULL       | NULL   | NULL  | NULL    | NULL   |
+--------+-------------+-----------+------+----------+------+-----------+------------+--------+-------+---------+--------+
8 rows in set (0.00 sec)
```

本示例 SQL 查询语句使用了 left join 关键字来连接 dept 和 emp 表，并以 deptno 字段为连接条件，查询并返回两张表中相关记录的字段值。具体地说，该查询会遍历 dept 表中的每一行记录，然后查找与之对应的 emp 表中的记录，如果两者中存在符合连接关系的记录，则会将它们的字段值合并为一条查询结果，并以列的形式呈现在最终的查询结果中。如果某个部门在 emp 表中没有关联记录，则该部门在查询结果中也会被保留，但其关联字段值会被填充为 null。

例 18.44 查询每个部门的员工的数量，示例 SQL 语句及其执行结果如下：

```
mysql> select dept.deptno,count(emp.deptno) from dept left join emp on dept.deptno
= emp.deptno group by dept.deptno;
+--------+-------------------+
| deptno | count(emp.deptno) |
```

```
+--------+--------------------+
|   10   |          2         |
|   20   |          1         |
|   30   |          2         |
|   40   |          2         |
|   50   |          0         |
+--------+--------------------+
5 rows in set (0.00 sec)
```

本示例 SQL 查询语句使用左连接（left join）的方式连接 dept 和 emp 表，并以 deptno 字段为连接条件，查询每个部门中的员工数量，然后使用 group by 子句将结果集按照 dept.deptno 列进行分组。需要注意的是，由于使用了左连接方式，因此即使某个部门在 emp 表中没有符合连接条件的记录，该部门在查询结果中也会出现，并且其表示员工数量的字段的值会被填充为 0，而不是 null。相对地，如果没有使用 group by 语法，则会返回所有符合连接条件的记录，其中包括没有员工的部门和在多个部门工作的员工。

18.7 子查询、多表查询

本节演示 MySQL 子查询、多表查询的实际操作。

例 18.45 列出与 lisi 从事相同职位的所有员工，显示姓名、职位、部门编号，示例 SQL 语句及其执行结果如下：

```
mysql> select ename, epost , deptno from emp where epost =(select epost from emp where ename='lisi');
+----------+--------+--------+
| ename    | epost  | deptno |
+----------+--------+--------+
| wanger   | worker |    10  |
| libai    | worker |    10  |
| lisi     | worker |    20  |
| litaibai | worker |    30  |
| zhoulei  | worker |    40  |
| xiaoli   | worker |    40  |
+----------+--------+--------+
6 rows in set (0.00 sec)
```

本示例 SQL 查询语句使用了子查询（subquery）的方式，在 emp 表中查询所有职位（epost）和部门编号（deptno）与指定员工（ename='lisi'，即名字为 lisi 的员工）相同的记录，并返回这些记录的姓名（ename）、职位和部门编号。需要注意的是，由于子查询指定了一个具体的员工名字，因此该查询只返回与该员工职位和部门编号相同的其他员工的信息，而不会返回所有具有相同职位和部门编号的员工记录。

例 18.46 列出薪资比部门编号为 30（销售部）的所有员工薪资都高的员工信息，显示员工姓名、薪资和部门名称，示例 SQL 语句及其执行结果如下：

```
# 外连接查询,查询所有员工、员工薪资和员工对应的部门名称
mysql>    select emp.ename, salary , dept.dname from emp left join dept on emp.deptno =dept.deptno ;
+----------+--------+------------+
| ename    | salary | dname      |
+----------+--------+------------+
| wanger   |   8000 | accounting |
| libai    |   4500 | accounting |
| lisi     |   6500 | research   |
| litaibai |   6000 | sales      |
| wangming |   7500 | sales      |
| zhoulei  |   5000 | operations |
| xiaoli   |   4800 | operations |
+----------+--------+------------+
7 rows in set (0.00 sec)

# 假设销售部门的最高薪资为 3000,列出薪资比 3000 高的员工信息
mysql>    select emp.ename,salary,dept.dname from emp left join dept on emp.deptno =dept .deptno where salary >3000;
+----------+--------+------------+
| ename    | salary | dname      |
+----------+--------+------------+
| wanger   |   8000 | accounting |
| libai    |   4500 | accounting |
| lisi     |   6500 | research   |
| litaibai |   6000 | sales      |
| wangming |   7500 | sales      |
| zhoulei  |   5000 | operations |
| xiaoli   |   4800 | operations |
+----------+--------+------------+
7 rows in set (0.00 sec)

# 求出销售部门的最高薪资
mysql>    select max(salary) from emp where deptno =30;
+-------------+
| max(salary) |
+-------------+
|        7500 |
+-------------+
1 row in set (0.00 sec)
# 合并两条查询 SQL
mysql>    select emp.ename, salary, dept.dname  from emp left join dept on emp.deptno =dept.deptno   where salary>(select max(salary) from emp where deptno =30);
+--------+--------+------------+
| ename  | salary | dname      |
+--------+--------+------------+
| wanger |   8000 | accounting |
+--------+--------+------------+
1 row in set (0.00 sec)
```

查询结果符合预期。

例 18.47 列出最低薪资大于 6500 的各种职位,显示职位和该职位的最低薪资,示例 SQL 语句及其执行结果如下:

```
mysql> select epost ,min(salary) from emp group by epost having min(salary )>6500;
+---------+-------------+
| epost   | min(salary) |
+---------+-------------+
| manager |        7500 |
+---------+-------------+
1 row in set (0.00 sec)
```

查询结果符合预期。

本示例 SQL 查询语句使用 group by 子句来将结果集按照 epost 列进行分组，并使用 min() 函数来计算每组中的最低薪水值。然后，使用 having 子句来筛选出最低薪资大于 6500 的组。最后，查询结果按照 epost 升序排列。需要注意的是，在 group by 子句之后使用 having 子句可以对分组后的结果进行条件过滤，而在 where 子句中则不能使用聚合函数（如 min 函数）。

例 18.48 列出在每个部门就职的员工数量、平均薪资，显示部门编号、员工数量、平均薪资，示例 SQL 语句及其执行结果如下：

```
mysql>    select deptno, count(*), avg(salary)  from emp group by deptno;
+--------+----------+-------------+
| deptno | count(*) | avg(salary) |
+--------+----------+-------------+
|     10 |        2 |   6250.0000 |
|     20 |        1 |   6500.0000 |
|     30 |        2 |   6750.0000 |
|     40 |        2 |   4900.0000 |
+--------+----------+-------------+
4 rows in set (0.00 sec)
```

查询结果符合预期。

例 18.49 列出每个部门薪资最高的员工信息，显示部门编号、员工姓名、薪资，示例 SQL 语句及其执行结果如下：

```
# 查询 emp 表中所有员工的部门编号、姓名、薪资
 mysql> select deptno, ename, salary from emp;
+--------+----------+--------+
| deptno | ename    | salary |
+--------+----------+--------+
|     10 | wanger   |   8000 |
|     10 | libai    |   4500 |
|     20 | lisi     |   6500 |
|     30 | litaibai |   6000 |
|     30 | wangming |   7500 |
|     40 | zhoulei  |   5000 |
|     40 | xiaoli   |   4800 |
+--------+----------+--------+
7 rows in set (0.00 sec)
# 查询 emp 表中每个部门的最高薪资，显示部门编号、最高薪资
mysql> select deptno, max(salary) from emp  group by deptno;
+--------+-------------+
| deptno | max(salary) |
```

```
+--------+-------------+
|   10   |    8000     |
|   20   |    6500     |
|   30   |    7500     |
|   40   |    5000     |
+--------+-------------+
4 rows in set (0.00 sec)
```
第二次查询的结果作为一张临时表和第一次查询进行关联查询
```
mysql> select emp.deptno,emp.ename,emp.salary from emp,(select deptno , max(salary) maxsal from emp group by deptno) t1  where t1.deptno=emp.deptno  and emp.salary=t1.maxsal;
+--------+----------+--------+
| deptno | ename    | salary |
+--------+----------+--------+
|     10 | wanger   |   8000 |
|     20 | lisi     |   6500 |
|     30 | wangming |   7500 |
|     40 | zhoulei  |   5000 |
+--------+----------+--------+
4 rows in set (0.00 sec)
```

查询结果符合预期。

18.8　本章练习

项目实例概述如下：

假设我们正在为 A 市设计一个图书馆管理系统。需要完成如下功能：

1）唯一标识书籍

某些书籍可能具有相同的名称，因此书名不能用来唯一标识书籍。图书管理员将书称为 item，item 由国际标准书号（ISBN）标识。具有相同书名的不同书籍可以用不同 ISBN 号区分。

2）预约书

如果借阅人需要借阅的书已经借出，那么借阅人可以预约这些书。等书归还以后，图书管理员会通知那些等待时间最长的预约者。借阅人一次最多可以预约 4 本书。

3）确定书籍是否可供借阅

在任何时间，图书管理员都需要知道有多少本同样的书已经被借出（如 3 本《红楼梦》被借出），以及哪些书被预约。图书馆数据库中的每本书都应该有或长或短的摘要，每当借阅人查询书籍信息时，管理员能够查询到书籍的摘要。

4）注册新成员

要成为图书馆的读者，借阅人必须提供地址和电话。然后管理员发给借阅人一张已经编好号码的卡片，可以用机器读取上面的数据。该卡一年内有效。

未成年人也可以成为图书馆的读者，但在注册时需要其家长也是该图书馆的成员。因此，未成年借阅人所持卡只在与其相关成年人的卡片过期前有效。图书馆中只保留未成年人的姓名和生日。当未成年借阅人年满 18 岁后，图书馆管理员必须将其删除，然后再自动转换为成年借阅人。

5）借书

书籍可以借出 14 天。借阅人一次只允许借 4 本书。如果过期，图书馆会在 1 周后通知借阅人。当借阅人找到他们需要的书籍后，就到前台办理借阅手续。管理员用刷卡机刷卡，并在屏幕上显示借阅人的账号信息，包括姓名、地址、电话号码以及卡片过期时间。那些已经过期或即将过期的信息将高亮显示。

屏幕上还会显示这个人的拖欠款信息（如果有的话），包括书名、借出日期和到期日期。这些信息是很有用的。所有的拖欠款信息按照时间排序，即最长过期的欠款出现在最前面，最近过期的欠款出现在最后。那些已经过期或即将过期的借书信息将高亮显示。

如果这个读者的账号状态良好，那么图书管理员可以将这些书借给他。图书管理员扫描书背面的条形码（包含 ISBN 和复制号），在计算机屏幕上会显示书的 ISBN、书名和作者信息。

如果这本书不可借，那么系统会显示警告信息。

6）还书

还书时，图书管理员用扫描仪扫一下书背面的条形码，这本书的 ISBN、书名、作者以及借阅人号码、姓名和还书日期就都会在计算机屏幕上显示出来。

偶尔也会发生这种情况，在借阅人还书时，图书管理员忘记更新该书的归还信息，那么，该书在数据库中的状态一直为已经借出。图书管理员应能够查阅该书的借阅人姓名、借书日期和还书日期。如果一个借阅人在书架上找到这本书，那么当他办理借阅时，计算机会显示出一条信息，提醒管理员该书已经借出。这时，管理员可以立即清除以前的借出信息来更新记录。

7）生成使用报表

有时，图书管理员需要搜集使用信息，为管理层提供参考。主要是一些关于流通量、各种书的受欢迎程度、归还可靠度以及平均借阅时间等信息。因此，图书管理员应该能够快速地汇总这些信息。下面是一些经常被问到的问题：

- 去年有多少借阅人？
- 最少借一本书的人占多大百分比？
- 借书最多的人，总共借了多少本书？
- 去年最少借出一次的书占多大百分比？
- 过期未还的书占多大百分比？
- 平均借阅时间是多少？
- 图书馆办理借阅的高峰时间是什么时候？

练 习

1. 根据如上信息，创建名为 mybook 的数据库，并根据如上要求列出需要的表结构，再通过 SQL 语句实现表的创建。
2. 插入测试数据。
3. 根据使用生成报表的要求，写出每一个报表相应的 SQL 语句。

附录 A

函数速查表

```
# 附录中会使用到的表结构和数据
mysql> create table userinfo (
    -> id int,
    -> name varchar (100),
    -> age int
    -> );
Query OK, 0 rows affected (0.03 sec)
mysql> insert into userinfo values(1,'zhangsan',18);
Query OK, 1 row affected (0.00 sec)
mysql> insert into userinfo values(2,'zhangsan2',19);
Query OK, 1 row affected (0.01 sec)
mysql> insert into userinfo values(3,'zhangsan3',20);
Query OK, 1 row affected (0.01 sec)
```

A.1　MySQL 字符串函数

A.1.1　ASCII(s)：返回字符串的 ASCII 码

返回字符串 s 的第一个字符的 ASCII 码。

示例：返回 name 字段第一个字母的 ASCII 码。

```
mysql> select ascii(name) from userinfo;
+-------------------+
| ascii(name)       |
+-------------------+
|    108            |
+-------------------+
1rows in set (0.01 sec)
```

A.1.2　char_length(s)：返回字符串长度

返回字符串 s 的长度。

示例：返回字符串"clay"的字符数。

```
mysql> select char_length("clay") as lengthofstring;
+----------------+
| lengthofstring |
+----------------+
|              4 |
+----------------+
1rows in set (0.01 sec)
```

A.1.3　character_length(s)：返回字符串长度

返回字符串 s 的长度。

示例：返回字符串"clay"的字符数。

```
mysql> select character_length("clay") as lengthofstring;
+----------------+
| lengthofstring |
+----------------+
|              4 |
+----------------+
1 row in set (0.00 sec)
```

A.1.4　concat(s1,s2,…,sn)：字符串合并

将 s1、s2 等多个字符串合并为一个字符串。

示例：合并多个字符串。

```
mysql> select concat("sql", "nosql", "newsql ", "end") as concatenatedstring;
+--------------------+
| concatenatedstring |
+--------------------+
| sql nosqlnewsql end |
+--------------------+
1 row in set (0.00 sec)
```

A.1.5　concat_ws(x, s1,s2,…,sn)：字符串合并

将 s1、s2 等多个字符串合并为一个字符串。与 concat(s1,s2,…)函数不同的是，concat_ws(x, s1,s2,…,sn)的每个字符串之间要加上 x，x 可以是分隔符。

示例：合并多个字符串，并添加分隔符。

```
mysql> select concat_ws("-", "sql", "nosql", "newsql")as concatenatedstring;
```

```
+--------------------+
| concatenatedstring |
+--------------------+
| sql-nosql-newsql   |
+--------------------+
1 row in set (0.00 sec)
```

A.1.6　field(s,s1,s2,…)：返回字符串出现的位置

返回第一个字符串 s 在字符串列表(s1,s2,…)中的位置。

示例：返回字符串"c"在列表中的位置。

```
mysql> select field("c", "a", "b", "c", "d");
+-------------------------------+
| field("c", "a", "b", "c", "d") |
+-------------------------------+
|                             3 |
+-------------------------------+
1 row in set (0.00 sec)
```

A.1.7　find_in_set(s1,s2)：返回字符串的匹配位置

返回在字符串 s2 中与 s1 匹配的字符串的位置。

示例：返回字符串"c"在指定字符串中的位置。

```
mysql> select find_in_set("c", "a,b,c,d");
+----------------------------+
| find_in_set("c", "a,b,c,d") |
+----------------------------+
|                          3 |
+----------------------------+
1 row in set (0.00 sec)
```

A.1.8　format(x,n)：数字格式化

可以将数字 x 格式化为"#,###.##"，将 x 保留到小数点后 n 位，最后一位四舍五入。

示例：格式化数字为"#,###.##"形式。

```
mysql> select format(6500700.52341, 2);
+-------------------------+
| format(6500700.52341, 2) |
+-------------------------+
| 6,500,700.52            |
+-------------------------+
1 row in set (0.00 sec)
```

A.1.9　insert(s1,x,len,s2)：替换字符串

用字符串 s2 替换 s1 中 x 位置开始长度为 len 的字符串。

示例：从字符串第一个位置开始的 4 个字符替换为"clay"。

```
mysql> select insert("this is aaaaaaaaa", 1, 4, "clay");
+-------------------------------------------+
| insert("this is aaaaaaaaa", 1, 4, "clay") |
+-------------------------------------------+
| clay is aaaaaaaaa                         |
+-------------------------------------------+
1 row in set (0.00 sec)
```

A.1.10　locate(s1,s)：获取字符串出现的位置

从字符串 s 中获取 s1 的开始位置。

示例：获取"clay"在字符串"this is clay"中的位置。

```
mysql> select locate("clay","this is clay");
+-------------------------------+
| locate("clay","this is clay") |
+-------------------------------+
|                             9 |
+-------------------------------+
1 row in set (0.00 sec)
```

A.1.11　lcase(s)：把字符串中的所有字母转换为小写字母

将字符串 s 中的所有字母转换为小写字母。

示例：将字符串"CLAY"中的所有字母转换为小写字母。

```
mysql> select lcase("clay");
+---------------+
| lcase("clay") |
+---------------+
| clay          |
+---------------+
1 row in set (0.00 sec)
```

A.1.12　left(s,n)：返回字符串的前 n 个字符

返回字符串 s 的前 n 个字符。

示例：返回字符串"clay"的前 2 个字符。

```
mysql> select left("clay",2);
+----------------+
| left("clay",2) |
```

```
+---------------+
| cl            |
+---------------+
1 row in set (0.00 sec)
```

A.1.13　lower(s)：把字符串中的所有字母转换为小写字母

将字符串 s 中的所有字母转换为小写字母。

示例：将字符串"CLAY"中的所有字母转换为小写字母。

```
mysql> select lower("CLAY");
+---------------+
| LOWER("CLAY") |
+---------------+
| clay          |
+---------------+
1 row in set (0.00 sec)
```

A.1.14　lpad(s1,len,s2)：字符串填充

在字符串 s1 的开始处填充字符串 s2，使字符串的长度达到 len。

示例：将字符串"ABC"填充到字符串"abc"的开始处，使字符串的长度达到 8。

```
mysql> select lpad("abc",8,"ABC");
+---------------------+
| LPAD("abc",8,"ABC") |
+---------------------+
| ABCABabc            |
+---------------------+
1 row in set (0.00 sec)
```

A.1.15　ltrim(s)：去掉字符串开始处的空格

去掉字符串 s 开始处的空格。

示例：去掉字符串"   this is clay"开始处的空格。

```
mysql> select ltrim("    this is clay") as trimmedstring;
+---------------+
| trimmedstring |
+---------------+
| this is clay  |
+---------------+
1 row in set (0.00 sec)
```

A.1.16　mid(s,n,len)：从字符串的指定位置截取子字符串

从字符串 s 的 n 位置截取长度为 len 的子字符串。

示例：从字符串"this is clay"中的第 1 个位置开始截取长度为 4 的子字符串。

```
mysql> select mid("this is clay", 1, 4);
+---------------------------+
| mid("this is clay", 1, 4) |
+---------------------------+
| this                      |
+---------------------------+
1 row in set (0.00 sec)
```

A.1.17　replace(s,s1,s2)：字符串替换

用字符串 s2 替换字符串 s 中的字符串 s1。

示例：将字符串"abcd"中的字符串"a"替换为字符串"x"。

```
mysql> select replace("abcd","a","x");
+-------------------------+
| replace("abcd","a","x") |
+-------------------------+
| xbcd                    |
+-------------------------+
1 row in set (0.00 sec
```

A.1.18　reverse(s)：字符串反转

将字符串 s 的顺序反过来。

示例：将字符串"clay"的顺序反过来。

```
mysql> select reverse("clay");
+-----------------+
| reverse("clay") |
+-----------------+
| yalc            |
+-----------------+
1 row in set (0.00 sec)
```

A.1.19　right(s,n)：返回字符串后 n 个字符

返回字符串 s 的后 n 个字符。

示例：返回字符串"clay"的后 3 个字符。

```
mysql> select right("clay",3);
+-----------------+
| right("clay",3) |
+-----------------+
| lay             |
+-----------------+
1 row in set (0.00 sec)
```

A.1.20　rpad(s1,len,s2)：在字符串中填充相应字符串

在字符串 s1 的结尾处添加字符串 s2，使字符串的长度达到 len。

示例：将字符串"clay"填充到"abc"字符串的结尾处，使字符串的长度达到 8。

```
mysql> select rpad("abc",8,"clay") ;
+----------------------+
| rpad("abc",8,"clay") |
+----------------------+
| abcclayc             |
+----------------------+
1 row in set (0.00 sec)
```

A.1.21　position(s1 in s)：获取子字符串在字符串中出现的位置

从字符串 s 中获取子字符串 s1 出现的位置。

示例：返回字符串"abc"中子字符串"a"的位置。

```
mysql> select position("a" in "abc");
+------------------------+
| position("a" in "abc") |
+------------------------+
|                      1 |
+------------------------+
1 row in set (0.00 sec)
```

A.1.22　repeat(s,n)：字符串重复拼接

将字符串 s 重复 n 次。

示例：将字符串"clay"重复两次。

```
mysql> select repeat("clay",2);
+------------------+
| repeat("clay",2) |
+------------------+
| clayclay         |
+------------------+
1 row in set (0.00 sec)
```

A.1.23　substring_index(s, s2, number)：返回字符串中第 n 个出现的子字符串

返回在字符串 s 中的第 number 个出现的分隔符 s2 之后的子字符串。如果 number 是正数，则返回从左边数第 number 个出现的分隔符 s2 之后的字符串；如果 number 是负数，则返回从右边数第 number 个分隔符 s2 之后的字符串。

示例 1：返回字符串"babacdd"中第 1 个出现的分隔符'a'之后的子字符串。

```
mysql> select substring_index("babacdd",'a',1);
+----------------------------------+
| substring_index("babacdd",'a',1) |
+----------------------------------+
| b                                |
+----------------------------------+
1 row in set (0.00 sec)
```

示例 2：返回字符串"babacdd"从右边数第 1 个出现的分隔符'a'之后的子字符串。

```
mysql> select substring_index("babacdd",'a',-1);
+-----------------------------------+
| substring_index("babacdd",'a',-1) |
+-----------------------------------+
| cdd                               |
+-----------------------------------+
1 row in set (0.00 sec)
```

A.1.24　trim(s)：去掉字符串开始和结尾处的空格

去掉字符串 s 开始和结尾处的空格。

示例：去掉字符串"   clay   "首尾的空格。

```
mysql> select trim("    clay    ") ;
+----------------------+
| trim("    clay    ") |
+----------------------+
| clay                 |
+----------------------+
1 row in set (0.00 sec)
```

A.1.25　ucase(s)：把字符串中的所有字母转换为大写字母

把字符串中的所有字母转换为大写字母。

示例：把字符串"clay"中的所有字母转换为大写字母。

```
mysql> select ucase("clay");
+---------------+
| ucase("clay") |
+---------------+
| CLAY          |
+---------------+
1 row in set (0.00 sec)
```

A.1.26　rtrim(s)：去掉字符串结尾处的空格

去掉字符串 s 结尾处的空格。

示例：去掉字符串"clay   "结尾处的空格。

```
mysql> select rtrim("clay    ") ;
+-------------------+
| rtrim("clay    ") |
+-------------------+
| clay              |
+-------------------+
1 row in set (0.00 sec)
```

A.1.27　strcmp(s1,s2)：比较字符串的大小

比较字符串 s1 和 s2，如果 s1 与 s2 相等，则返回 0；如果 s1>s2，则返回 1；如果 s1<s2，则返回–1。

示例：比较字符串"abc"和 "bcd"的大小。

```
mysql> select strcmp("abc", "bcd");
+----------------------+
| strcmp("abc", "bcd") |
+----------------------+
|                   -1 |
+----------------------+
1 row in set (0.00 sec)
```

A.1.28　substr(s, start, length)：从字符串中截取子字符串

从字符串 s 的 start 位置开始截取长度为 length 的子字符串。

示例：从字符串"this is clay"中的第 3 个位置开始截取长度为 2 的子字符串。

```
mysql> select substr("this is clay", 3, 2);
+------------------------------+
| substr("this is clay", 3, 2) |
+------------------------------+
| IS                           |
+------------------------------+
1 row in set (0.00 sec)
```

A.1.29　substring(s, start, length)：从字符串中截取子字符串

从字符串 s 的 start 位置开始截取长度为 length 的子字符串。

示例：从字符串"this is clay"中的第 1 个位置开始截取长度为 3 的子字符串。

```
mysql> select substring("this is clay", 1, 3) ;
+---------------------------------+
| substring("this is clay", 1, 3) |
+---------------------------------+
| thi                             |
+---------------------------------+
```

```
1 row in set (0.01 sec)
```

A.1.30　md5(s)：字符串加密

对字符串 s 进行加密。

示例：对字符串"password"进行 MD5 加密。

```
mysql> select md5("password");
+----------------------------------+
| md5("password")                  |
+----------------------------------+
| 319f4d26e3c536b5dd871bb2c52e3178 |
+----------------------------------+
1 row in set (0.00 sec)
```

A.1.31　inet_aton (ip)：把 IP 地址转换为数字

把 IP 地址转换为数字。

示例：把 IP 地址 192.168.1.201 转换为数字。

```
mysql> select inet_aton("192.168.1.201");
+----------------------------+
| inet_aton("192.168.1.201") |
+----------------------------+
|                 3232235977 |
+----------------------------+
1 row in set (0.00 sec)
```

A.1.32　inet_ntoa(s)：把数字转换为 IP 地址

把数字 s 转换为 IP 地址。

示例：把 3232235977 转换为 IP 地址。

```
mysql> select inet_ntoa(3232235977);
+-----------------------+
| inet_ntoa(3232235977) |
+-----------------------+
| 192.168.1.201         |
+-----------------------+
1 row in set (0.00 sec)
```

A.2 MySQL 数字函数

A.2.1 abs(x)：求绝对值

返回 x 的绝对值。

示例：返回–6 的绝对值。

```
mysql> select abs(-6);
+---------+
| abs(-6) |
+---------+
|       6 |
+---------+
1 row in set (0.00 sec)
```

A.2.2 sign(x)：返回数字符号

返回 x 的符号，x 是负数、0、正数时，分别返回–1、0 和 1。

示例 1：返回–5 的符号。

```
mysql> select sign(-5);
+----------+
| sign(-5) |
+----------+
|       -1 |
+----------+
1 row in set (0.00 sec)
```

示例 2：返回 5 的符号。

```
mysql> select sign(5);
+---------+
| sign(5) |
+---------+
|       1 |
+---------+
1 row in set (0.00 sec)
```

A.2.3 acos(x)：求反余弦值

求 x 的反余弦值。

示例：求弧度 0.15 的反余弦值。

```
mysql> select acos(0.15);
+--------------------+
| acos(0.15)         |
+--------------------+
```

```
| 1.4202280540182106 |
+--------------------+
1 row in set (0.00 sec)
```

A.2.4　asin(x)：求反正弦值

求 x 的反正弦值。

示例：求弧度 0.15 的反正弦值。

```
mysql> select asin(0.15);
+--------------------+
| asin(0.15)         |
+--------------------+
| 0.15056827277668602 |
+--------------------+
1 row in set (0.00 sec)
```

A.2.5　atan (x)：求反正切值

求 x 的反正切值。

示例：求弧度 1.5 的反正切值。

```
mysql> select atan(1.5);
+-------------------+
| atan(1.5)         |
+-------------------+
| 0.982793723247329 |
+-------------------+
1 row in set (0.00 sec)
```

A.2.6　sin(x)：求正弦值

求 x 的正弦值。

示例：求弧度 0.5 的正弦值。

```
mysql> select sin(0.5);
+-------------------+
| sin(0.5)          |
+-------------------+
| 0.479425538604203 |
+-------------------+
1 row in set (0.00 sec)
```

A.2.7　cos(x)：求余弦值

求 x 的余弦值。

示例：求弧度 2 的余弦值。

```
mysql> select cos(2);
+--------------------+
| cos(2)             |
+--------------------+
| -0.4161468365471424 |
+--------------------+
1 row in set (0.00 sec)
```

A.2.8　cot(x)：求余切值

求 x 的余切值。

示例：求弧度 2 的余切值。

```
mysql> select cot(2);
+---------------------+
| cot(2)              |
+---------------------+
| -0.45765755436028577 |
+---------------------+
1 row in set (0.00 sec)
```

A.2.9　tan(x)：求正切值

求 x 的正切值。

示例：求弧度 2.5 的正切值。

```
mysql> select tan(2.5);
+--------------------+
| tan(2.5)           |
+--------------------+
| -0.7470222972386603 |
+--------------------+
1 row in set (0.00 sec)
```

A.2.10　degrees(x)：将弧度转换为角度

将弧度 x 转换为角度。

示例：将弧度 3.14 转换为角度。

```
mysql> select degrees(3.14);
+------------------+
| degrees(3.14)    |
+------------------+
| 179.9087476710785 |
+------------------+
1 row in set (0.00 sec)
```

A.2.11 radians(x)：将角度转换为弧度

将角度 x 转换为弧度。

示例：将角度 180° 转换为弧度。

```
mysql> select radians(180);
+-------------------+
| radians(180)      |
+-------------------+
| 3.141592653589793 |
+-------------------+
1 row in set (0.00 sec)
```

A.2.12 exp(x)：返回 e 的 x 次方

返回 e 的 x 次方。

示例：计算 e 的 4 次方。

```
mysql> select exp(4);
+--------------------+
| exp(4)             |
+--------------------+
| 54.598150033144236 |
+--------------------+
1 row in set (0.00 sec)
```

A.2.13 ceil(x)：返回不小于 x 的最小整数

返回不小于 x 的最小整数。

示例：返回大于或等于 2.5 的最小整数。

```
mysql> select ceil(2.5).
+-----------+
| ceil(2.5) |
+-----------+
|         3 |
+-----------+
1 row in set (0.00 sec)
```

A.2.14 ceiling(x)：返回不小于 x 的最小整数

返回不小于 x 的最小整数。

示例：返回大于或等于 2.5 的最小整数。

```
mysql> select ceiling(2.5);
+--------------+
| ceiling(2.5) |
```

```
+-------------+
|           3 |
+-------------+
1 row in set (0.00 sec)
```

A.2.15 floor(x)：返回不大于 x 的最大整数

返回小于或等于 x 的最大整数。

示例：返回小于或等于 1.5 的最大整数。

```
mysql> select floor(1.5);
+------------+
| floor(1.5) |
+------------+
|          1 |
+------------+
1 row in set (0.00 sec)
```

A.2.16 round(x)：返回最接近 x 的整数

返回最接近 x 的整数。

示例 1：返回最接近 12.23 的整数。

```
mysql> select round(12.23) ;
+--------------+
| round(12.23) |
+--------------+
|           12 |
+--------------+
1 row in set (0.00 sec)
```

示例 2：返回最接近 12.93 的整数。

```
mysql> select round(12.93) ;
+--------------+
| round(12.93) |
+--------------+
|           13 |
+--------------+
1 row in set (0.00 sec)
```

A.2.17 greatest(expr1,expr2, …)：返回列表中的最大值

返回列表中的最大值。

示例：返回列表（13, 16, 38, 18, 25）中的最大值。

```
mysql> select greatest(13, 16, 38, 18, 25);
+------------------------------+
| greatest(13, 16, 34, 18, 25) |
```

```
+----------------------------+
|                         38 |
+----------------------------+
1 row in set (0.00 sec)
```

A.2.18　least(expr1, expr2, expr3, …)：返回列表中的最小值

返回列表中的最小值。

示例：返回列表（13, 16, 38, 18, 25）中的最小值。

```
mysql> select least(13, 16, 38, 18, 25);
+--------------------------+
| least(13, 16, 38, 18, 25) |
+--------------------------+
|                       13 |
+--------------------------+
1 row in set (0.00 sec)
```

A.2.19　ln(x)：求自然对数

返回 x 的自然对数。

示例：返回以 e 为底的 4 的自然对数值。

```
mysql> select ln(4);
+--------------------+
| ln(4)              |
+--------------------+
| 1.3862943611198906 |
+--------------------+
1 row in set (0.00 sec)
```

A.2.20　log(x)或log(base, x)：求对数

返回 x 的自然对数（以 e 为底的对数），如果带有 base 参数，则返回以 base 为底的对数。

示例 1：返回以 e 为底的 1.3862943611198906 的对数值。

```
mysql> select log(1.3862943611198906);
+-------------------------+
| log(1.3862943611198906) |
+-------------------------+
|      0.32663425997828094 |
+-------------------------+
1 row in set (0.00 sec)
```

示例 2：返回以 5 为底的 2 的对数值。

```
mysql> select log(2,5) ;
+------------------+
| log(2,5)         |
```

```
+------------------+
| 2.321928094887362 |
+------------------+
1 row in set (0.00 sec)
```

A.2.21 log10(x)：求以 10 为底的对数

返回以 10 为底的对数。

示例：返回以 10 为底的 1000 的对数值。

```
mysql> select log10(1000);
+-------------+
| log10(1000) |
+-------------+
|           3 |
+-------------+
1 row in set (0.00 sec)
```

A.2.22 log2(x)：求以 2 为底的对数

返回以 2 为底的对数。

示例：返回以 2 为底的 8 的对数值。

```
mysql> select log2(8);
+---------+
| log2(8) |
+---------+
|       3 |
+---------+
1 row in set (0.00 sec)
```

A.2.23 max(expression)：求最大值

返回字段 expression 中的最大值。

示例：返回数据表 userinfo 中字段 age 的最大值。

```
mysql> select age from userinfo;
+------+
| age  |
+------+
|   18 |
|   19 |
|   20 |
+------+
3 rows in set (0.00 sec)
mysql> select max(age) from userinfo;
+----------+
| max(age) |
+----------+
```

```
|       20 |
+----------+
1 row in set (0.00 sec)
```

A.2.24　min(expression)：求最小值

返回字段 expression 中的最小值。

示例：返回数据表 userinfo 中字段 age 的最小值。

```
mysql> select age from userinfo;
+------+
| age  |
+------+
|   18 |
|   19 |
|   20 |
+------+
3 rows in set (0.00 sec)
mysql> select min(age) from userinfo;
+----------+
| min(age) |
+----------+
|       18 |
+----------+
1 row in set (0.00 sec)
```

A.2.25　sum(expression)：求总和

返回指定字段的总和。

示例：计算数据表 userinfo 中字段 age 的总和。

```
mysql> select age from userinfo;
+------+
| age  |
+------+
|   18 |
|   19 |
|   20 |
+------+
3 rows in set (0.00 sec)
mysql> select sum(age) from userinfo;
+----------+
| sum(age) |
+----------+
|       57 |
+----------+
1 row in set (0.00 sec)
```

A.2.26　avg(expression)：求平均值

返回指定字段的平均值。

示例：返回数据表 userinfo 中 age 字段的平均值。

```
mysql> select age from userinfo;
+------+
| age  |
+------+
|   18 |
|   19 |
|   20 |
+------+
3 rows in set (0.00 sec)
mysql> select avg(age) from userinfo;
+----------+
| avg(age) |
+----------+
|       19 |
+----------+
1 row in set (0.00 sec)
```

A.2.27　count(expression)：求总记录数

返回查询记录的总数，expression 参数是一个字段或者"*"。

示例：返回数据表 userinfo 中总共有多少条记录。

```
mysql> select count(*) from userinfo;
+----------+
| count(*) |
+----------+
|        3 |
+----------+
1 row in set (0.00 sec)
mysql> select count(*) from userinfo where age>19;
+----------+
| count(*) |
+----------+
|        0 |
+----------+
1 row in set (0.00 sec)
```

A.2.28　mod(x,y)：求余数

返回 x 除以 y 的余数。

示例：返回 6 除以 4 的余数。

```
mysql> select mod(6,4);
+----------+
| mod(6,4) |
```

```
+----------+
|     2    |
+----------+
1 row in set (0.00 sec)
```

A.2.29　pow(x,y)：求 x 的 y 次方

返回 x 的 y 次方。

示例：返回 3 的 3 次方。

```
mysql> select pow(3,3);
+----------+
| pow(3,3) |
+----------+
|       27 |
+----------+
1 row in set (0.00 sec)
```

A.2.30　sqrt(x)：求平方根

返回 x 的平方根。

示例：求 25 的平方根。

```
mysql> select sqrt(25);
+----------+
| sqrt(25) |
+----------+
|        5 |
+----------+
1 row in set (0.00 sec)
```

A.2.31　rand()：求随机数

返回 0～1 的随机数。

示例：返回 0～1 的随机数。

```
mysql> select rand();
+--------------------+
| rand()             |
+--------------------+
| 0.9472021376222391 |
+--------------------+
1 row in set (0.00 sec)
mysql> select rand();
+--------------------+
| rand()             |
+--------------------+
| 0.6028625579577522 |
+--------------------+
1 row in set (0.00 sec)
```

A.2.32　truncate(x,y)：返回保留到小数点后 n 位的值

返回数值 x 保留到小数点后 y 位的值（与 round 函数最大的区别是不会进行四舍五入）。

示例：求 1.8578 保留到小数点后 2 位的值。

```
mysql> select truncate(1.8578,2);
+--------------------+
| truncate(1.8578,2) |
+--------------------+
|               1.85 |
+--------------------+
1 row in set (0.00 sec)
```

A.3　MySQL 日期函数

A.3.1　adddate(d,n)：返回指定日期加上指定天数后的日期

计算起始日期 d 加上 n 天后的日期。

示例：计算在 2017-06-15 的基础上加上 60 天后的日期。

```
mysql> select adddate("2017-06-15", 60);
+---------------------------+
| adddate("2017-06-15", 60) |
+---------------------------+
| 2017-08-14                |
+---------------------------+
1 row in set (0.00 sec)
```

A.3.2　addtime(t,n)：返回指定时间加上指定时间后的时间

返回时间 t 加上时间 n 后的时间（是一个时间表达式）。

示例 1：返回 2021-06-06 23:23:10 加 8 秒的时间。

```
mysql> select addtime("2021-06-06 23:23:10", 8);
+-----------------------------------+
| addtime("2021-06-06 23:23:10", 8) |
+-----------------------------------+
| 2021-06-06 23:23:18               |
+-----------------------------------+
1 row in set (0.00 sec)
```

示例 2：返回 2021-06-06 23:23:10 加 1 小时 10 分 5 秒的时间。

```
mysql> select addtime("2021-06-06 23:23:10", "1:10:5");
+------------------------------------------+
| addtime("2021-06-06 23:23:10", "1:10:5") |
+------------------------------------------+
```

```
| 2021-06-07 00:33:15                      |
+------------------------------------------+
1 row in set (0.00 sec)
```

A.3.3　curdate()：返回当前日期

返回当前日期。

示例：返回当前日期。

```
mysql> select curdate();
+------------+
| curdate()  |
+------------+
| 2021-06-06 |
+------------+
1 row in set (0.00 sec)
```

A.3.4　current_date()：返回当前日期

返回当前日期。

示例：返回当前日期。

```
mysql> select current_date();
+----------------+
| current_date() |
+----------------+
| 2021-06-06     |
+----------------+
1 row in set (0.00 sec)
```

A.3.5　current_time()：返回当前时间

返回当前时间。

示例：返回当前时间。

```
mysql> select current_time();
+----------------+
| current_time() |
+----------------+
| 15:27:25       |
+----------------+
1 row in set (0.00 sec)
```

A.3.6　current_timestamp()：返回当前日期和时间

返回当前日期和时间。

示例：返回当前日期和时间。

```
mysql> select current_timestamp();
+---------------------+
| current_timestamp() |
+---------------------+
| 2021-06-06 15:36:47 |
+---------------------+
1 row in set (0.00 sec)
```

A.3.7　curtime()：返回当前时间

返回当前时间。

示例：返回当前时间。

```
mysql> select curtime();
+-----------+
| curtime() |
+-----------+
| 15:38:59  |
+-----------+
1 row in set (0.00 sec)
```

A.3.8　date(t)：从指定日期时间中提取日期值

从指定日期时间中提取日期值。

示例：从 2021-06-16 23:11:11 中提取日期值。

```
mysql> select date("2021-06-16 23:11:11");
+-----------------------------+
| date("2021-06-16 23:11:11") |
+-----------------------------+
| 2021-06-16                  |
+-----------------------------+
1 row in set (0.00 sec)
```

A.3.9　datediff(d1,d2)：返回两个日期相隔的天数

计算日期 d1 和 d2 之间相隔的天数。

示例：计算日期 2021-08-08 和 2021-06-06 之间相隔的天数。

```
mysql> select datediff("2021-08-08","2021-06-06");
+-------------------------------------+
| datediff("2021-08-08","2021-06-06") |
+-------------------------------------+
|                                  63 |
+-------------------------------------+
1 row in set (0.00 sec)
```

A.3.10 date_add(d, interval expr type)：给指定日期加上一个时间段后的日期

计算起始日期 d 加上一个时间段后的日期。

示例 1：计算起始日期 2021-11-11 11:11:11 加 1 天后的日期。

```
mysql> select adddate("2021-11-11 11:11:11",1);
+----------------------------------+
| adddate("2021-11-11 11:11:11",1) |
+----------------------------------+
| 2021-11-12 11:11:11              |
+----------------------------------+
1 row in set (0.00 sec)
```

示例 2：计算起始日期 2021-11-11 11:11:11 加 6 分钟后的日期。

```
mysql> select adddate("2021-11-11 11:11:11", interval 6 minute);
+---------------------------------------------------+
| adddate("2021-11-11 11:11:11", interval 6 minute) |
+---------------------------------------------------+
| 2021-11-11 11:17:11                               |
+---------------------------------------------------+
1 row in set (0.00 sec)
```

A.3.11 date_format(d,f)：根据表达式显示日期

按表达式 f 的要求显示日期 d。

示例：按表达式%Y-%m-%d %r 的要求显示日期 2021-11-11 11:11:11。

```
mysql> select date_format("2021-11-11 11:11:11","%y-%m-%d %r");
+--------------------------------------------------+
| date_format("2021-11-11 11:11:11","%y-%m-%d %r") |
+--------------------------------------------------+
| 2021-11-11 11:11:11 AM                           |
+--------------------------------------------------+
1 row in set (0.00 sec)
```

A.3.12 date_sub(date, interval expr type)：从当前日期减去指定的时间间隔

从当前日期减去指定的时间间隔。

示例：从当前日期减去 2 天。

```
mysql> select date_sub(now(),interval 2 day) ;
+--------------------------------+
| date_sub(now(),interval 2 day) |
+--------------------------------+
| 2021-06-04 15:55:09            |
```

```
+------------------------------+
1 row in set (0.00 sec)
```

A.3.13　day(d)：返回日期值中的日值

返回日期值 d 中的日值。

示例：返回日期值 2022-06-07 中的日值。

```
mysql> select day("2022-06-07");
+-------------------+
| day("2022-06-07") |
+-------------------+
|                 7 |
+-------------------+
1 row in set (0.00 sec)
```

A.3.14　dayname(d)：返回指定日期是星期几

返回日期 d 是星期几。

示例：返回日期 2021-06-06 是星期几。

```
mysql> select dayname("2021-06-06");
+-----------------------+
| dayname("2021-06-06") |
+-----------------------+
| Sunday                |
+-----------------------+
1 row in set (0.00 sec)
```

A.3.15　dayofmonth(d)：返回指定日期是本月的第几天

计算日期 d 是本月的第几天。

示例：计算日期 2021-06-06 是本月的第几天。

```
mysql> select dayofmonth("2021-06-06");
+--------------------------+
| dayofmonth("2021-06-06") |
+--------------------------+
|                        6 |
+--------------------------+
1 row in set (0.00 sec)
```

A.3.16　dayofweek(d)：返回指定日期是所在星期的第几天

返回日期 d 是所在星期的第几天（注意：一周的第一天为星期日，源于西方文化）。

示例：返回日期 2021-06-06 是所在星期的第几天。

```
mysql> select dayofweek("2021-06-06");
+-------------------------+
| dayofweek("2021-06-06") |
+-------------------------+
|                       1 |
+-------------------------+
1 row in set (0.00 sec)
```

A.3.17　dayofyear(d)：返回指定日期是本年的第几天

计算日期 d 是本年的第几天。

示例：计算日期 2021-06-06 是本年的第几天。

```
mysql> select dayofyear("2021-06-06");
+-------------------------+
| dayofyear("2021-06-06") |
+-------------------------+
|                     157 |
+-------------------------+
1 row in set (0.00 sec)
```

A.3.18　extract (type from d)：根据对应格式返回日期

从日期 d 中获取指定的值，type 指定返回的值。

type 可以取值为表 A-1 中的任一类型。

表 A-1　type 取值类型

取　　值	说　　明
hour	小时
minute	分
second	秒
microsecond	毫秒
year	年
month	月
day	日
week	周
quarter	季
year_month	年和月
day_hour	日和小时
day_minute	日和分钟
day_second	日和秒
hour_minute	小时和分
hour_second	小时和秒
minute_second	分钟和秒

示例 1：获取 2021-06-06 23:11:11 日期中的分钟部分。

```
mysql> select extract(minute from '2021-06-06 23:11:11');
+--------------------------------------------+
| extract(minute from "2021-06-06 23:11:11") |
+--------------------------------------------+
|                                         11 |
+--------------------------------------------+
1 row in set (0.00 sec)
```

示例 2：获取 2021-06-06 23:11:11 日期是当月的第几天。

```
mysql> select extract(day from '2021-06-06 23:11:11');
+-----------------------------------------+
| extract(day from "2021-06-06 23:11:11") |
+-----------------------------------------+
|                                       6 |
+-----------------------------------------+
1 row in set (0.00 sec)
```

示例 3：获取 2021-06-06 23:11:11 日期是当年的第几周。

```
mysql> select extract(week from "2021-06-06 23:11:11");
+------------------------------------------+
| extract(week from "2021-06-06 23:11:11") |
+------------------------------------------+
|                                       23 |
+------------------------------------------+
1 row in set (0.00 sec)
```

A.3.19　from_days(n)：返回元年加 n 天的日期

计算从 0000 年 1 月 1 日开始 n 天后的日期。

示例：计算从 0000 年 1 月 1 日开始 6000 天后的日期。

```
mysql> select from_days(6000);
+-----------------+
| from_days(6000) |
+-----------------+
| 0016-06-05      |
+-----------------+
1 row in set (0.00 sec)
```

A.3.20　hour(t)：返回指定时间中的小时数

返回 t 中的小时数。

示例：返回 2021-06-06 23:11:11 中的小时数。

```
mysql> select hour("2021-06-06 23:11:11");
+-----------------------------+
| hour("2021-06-06 23:11:11") |
```

```
+--------------------------------+
|                            23  |
+--------------------------------+
1 row in set (0.00 sec)
```

A.3.21　last_day(d)：返回给定日期所在月份的最后一天

返回给定日期所在月份的最后一天。

示例：返回 2021-06-06 23:11:11 那一月份的最后一天。

```
mysql> select last_day("2021-06-06 23:11:11");
+---------------------------------+
| last_day("2021-06-06 23:11:11") |
+---------------------------------+
| 2021-06-30                      |
+---------------------------------+
1 row in set (0.00 sec)
```

A.3.22　localtime()：返回当前日期和时间

返回当前日期和时间。

示例：返回当前日期和时间。

```
mysql> select localtime();
+---------------------+
| localtime()         |
+---------------------+
| 2021-06-06 16:20:54 |
+---------------------+
1 row in set (0.00 sec)
```

A.3.23　localtimestamp()：返回当前日期和时间

返回当前日期和时间。

示例：返回当前日期和时间。

```
mysql> select localtimestamp();
+---------------------+
| localtimestamp()    |
+---------------------+
| 2021-06-06 16:21:43 |
+---------------------+
1 row in set (0.00 sec)
```

A.3.24　makedate(year, day-of-year)：时间组合天数

基于给定参数年份（year）和所在年中的天数序号（day-of-year）返回一个日期。

示例：返回 2021 年的第 31 天的日期。

```
mysql> select makedate(2021, 31);
+--------------------+
| makedate(2021, 31) |
+--------------------+
| 2021-01-31         |
+--------------------+
1 row in set (0.00 sec)
```

A.3.25　maketime(hour, minute, second)：时间组合

组合时间，参数分别为小时、分钟、秒。

示例：返回组合参数分别为 1 小时、2 分钟、4 秒的时间。

```
mysql> select maketime(1, 2, 4);
+-------------------+
| maketime(1, 2, 4) |
+-------------------+
| 01:02:04          |
+-------------------+
1 row in set (0.00 sec)
```

A.3.26　microsecond(date)：返回时间参数中的微秒数

返回时间参数中的微秒数。

示例：返回 2021-06-20 09:34:00.0000213 参数中的微秒数。

```
mysql> select microsecond("2021-06-20 09:34:00.0000213");
+--------------------------------------------+
| microsecond("2021-06-20 09:34:00.0000213") |
+--------------------------------------------+
|                                         21 |
+--------------------------------------------+
1 row in set (0.00 sec)
```

A.3.27　minute(t)：返回指定时间中的分钟数

返回时间 t 中的分钟数。

示例：返回 2021-06-20 09:34:00.0000213 中的分钟值。

```
mysql> select minute("2021-06-20 09:34:00.0000213");
+---------------------------------------+
| minute("2021-06-20 09:34:00.0000213") |
+---------------------------------------+
|                                    34 |
+---------------------------------------+
1 row in set (0.00 sec)
```

A.3.28　monthname(d)：返回日期中的月份名称

返回日期中的月份名称。

示例：返回 2021-06-20 09:34:00.0000213 中的月份名称。

```
mysql> select monthname("2021-06-20 09:34:00.0000213");
+------------------------------------------+
| monthname("2021-06-20 09:34:00.0000213") |
+------------------------------------------+
| June                                     |
+------------------------------------------+
1 row in set (0.00 sec)
```

A.3.29　month(d)：返回日期中的月份值

返回日期 d 中的月份值。

示例：返回 2021-06-20 09:34:00 中的月份值。

```
mysql> select month("2021-06-20 09:34:00");
+------------------------------+
| month("2021-06-20 09:34:00") |
+------------------------------+
|                            6 |
+------------------------------+
1 row in set (0.00 sec)
```

A.3.30　now()：返回当前日期和时间

返回当前日期和时间。

示例：返回当前日期和时间。

```
mysql> select now();
+---------------------+
| now()               |
+---------------------+
| 2021-06-06 16:30:23 |
+---------------------+
1 row in set (0.00 sec)
```

A.3.31　period_add(period, number)：日期加月份数

为年-月组合的日期添加月份数。

示例：为日期 2021-03 添加 5 个月。

```
mysql> select period_add(202103, 5);
+-----------------------+
| period_add(202103, 5) |
+-----------------------+
|                202108 |
```

```
+---------------------+
1 row in set (0.01 sec)
```

A.3.32　period_diff(period1, period2)：返回两个时间之间的月份差

返回两个时间之间的月份差值。

示例：返回 202110 和 202103 之间的月份差值。
```
mysql> select period_diff(202110, 202103);
+-----------------------------+
| period_diff(202110, 202103) |
+-----------------------------+
|                           7 |
+-----------------------------+
1 row in set (0.00 sec)
```

A.3.33　quarter(d)：返回日期对应的季度数

返回日期 d 是第几季，范围是 1~4。

示例：返回日期 2021-06-07 是第几季。
```
mysql> select quarter("2021-06-07");
+-----------------------+
| quarter("2021-06-07") |
+-----------------------+
|                     2 |
+-----------------------+
1 row in set (0.00 sec)
```

A.3.34　second(t)：返回指定时间中的秒数

返回时间 t 中的秒数。

示例：返回 2021-06-07 11:2:3 中的秒钟值。
```
mysql> select second("2021-06-07 11:2:3");
+-----------------------------+
| second("2021-06-07 11:2:3") |
+-----------------------------+
|                           3 |
+-----------------------------+
1 row in set (0.00 sec)
```

A.3.35　sec_to_time(s)：秒数转换为时间

将以秒为单位的时间 s 转换为时分秒的格式。

示例：将 5467 秒转换为时分秒的格式。
```
mysql> select sec_to_time(5467);
```

```
+-------------------+
| sec_to_time(5467) |
+-------------------+
| 01:31:07          |
+-------------------+
1 row in set (0.00 sec)
```

A.3.36　str_to_date(string, format_mask)：把字符串转换为日期

将字符串转换为日期。

示例：将字符串"August 9 2021"转换为日期。

```
mysql> select str_to_date("august 9 2021", "%m %d %y");
+------------------------------------------+
| str_to_date("august 9 2021", "%m %d %y") |
+------------------------------------------+
| 2021-08-09                               |
+------------------------------------------+
1 row in set (0.00 sec)
```

A.3.37　subdate(d,n)：从指定日期减去指定天数后的日期

从日期 d 减去 n 天后的日期。

示例：日期 2021-11-11 11:11:11 减去 2 天后的日期。

```
mysql> select subdate("2021-11-11 11:11:11",2);
+----------------------------------+
| subdate("2021-11-11 11:11:11",2) |
+----------------------------------+
| 2021-11-09 11:11:11              |
+----------------------------------+
1 row in set (0.00 sec)
```

A.3.38　subtime(t,n)：从指定时间减去指定秒数后的时间

从时间 t 减去 n 秒后的时间。

示例：从时间 2021-11-11 11:11:11 减去 6 秒后的时间。

```
mysql> select subtime("2021-11-11 11:11:11", 6);
+-----------------------------------+
| subtime("2021-11-11 11:11:11", 6) |
+-----------------------------------+
| 2021-11-11 11:11:05               |
+-----------------------------------+
1 row in set (0.00 sec)
```

A.3.39 sysdate()：返回当前日期和时间

返回当前日期和时间。

示例：返回当前日期和时间。

```
mysql> select sysdate();
+---------------------+
| sysdate()           |
+---------------------+
| 2021-06-06 16:42:45 |
+---------------------+
1 row in set (0.00 sec)
```

A.3.40 time(expression)：提取日期时间参数中的时间部分

提取日期时间参数中的时间部分。

示例：提取 2021-06-06 16:42:45 的时间部分。

```
mysql> select time("2021-06-06 16:42:45");
+-----------------------------+
| time("2021-06-06 16:42:45") |
+-----------------------------+
| 16:42:45                    |
+-----------------------------+
1 row in set (0.00 sec)
```

A.3.41 time_format(t,f)：根据表达式显示时间

按表达式 f 的要求显示时间 t。

示例：按表达式%r 的要求显示时间 2021-06-06 16:42:45。

```
mysql> select time_format("2021-06-06 16:42:45","%r");
+-----------------------------------------+
| time_format("2021-06-06 16:42:45","%r") |
+-----------------------------------------+
| 04:42:45 PM                             |
+-----------------------------------------+
1 row in set (0.00 sec)
```

A.3.42 time_to_sec(t)：把时间转换为秒数

将时间 t 转换为秒数。

示例：将时间 16:42:45 转换为秒数。

```
mysql> select time_to_sec("16:42:45");
+-------------------------+
| time_to_sec("16:42:45") |
```

```
+------------------------+
|                  60165 |
+------------------------+
1 row in set (0.00 sec)
```

A.3.43　timediff(time1, time2)：计算时间差

计算时间 time1 和 time2 的差值。

示例：计算时间 2021-06-06 16:42:45 和 2020-06-06 16:42:45 的差值。

```
mysql> select timediff("2021-06-06 16:42:45", "2020-06-06 16:42:45");
+--------------------------------------------------------+
| timediff("2021-06-06 16:42:45", "2020-06-06 16:42:45") |
+--------------------------------------------------------+
| 838:59:59                                              |
+--------------------------------------------------------+
1 row in set, 1 warning (0.00 sec)
```

A.3.44　timestamp(expression, interval)：指定时间加上时间间隔后的时间

指定时间加上时间间隔后的时间。

示例：时间 2021-06-06 16:42:45 和时间间隔 13:10:11 相加，求相加后的时间。

```
mysql> select timestamp("2021-06-06 16:42:45", "13:10:11");
+----------------------------------------------+
| timestamp("2021-06-06 16:42:45", "13:10:11") |
+----------------------------------------------+
| 2021-06-07 05:52:56                          |
+----------------------------------------------+
1 row in set (0.00 sec)
```

A.3.45　to_days(d)：计算元年到当前日期的天数

计算日期 d 距离 0000 年 1 月 1 日的天数。

示例：计算日期 2021-06-06 16:42:45 距离 0000 年 1 月 1 日的天数。

```
mysql> select to_days('2021-06-06 16:42:45');
+--------------------------------+
| to_days('2021-06-06 16:42:45') |
+--------------------------------+
|                         737947 |
+--------------------------------+
1 row in set (0.00 sec)
```

A.3.46　week(d)：返回指定日期是本年的第几周

计算日期 d 是本年的第几周，范围是 0~53。

示例：计算日期 2021-06-06 16:42:45 是本年的第几周，范围是 0~53。

```
mysql> select week('2021-06-06 16:42:45');
+-----------------------------+
| week('2021-06-06 16:42:45') |
+-----------------------------+
|                          22 |
+-----------------------------+
1 row in set (0.00 sec)
```

A.3.47　weekday(d)：返回指定日期是星期几

返回日期 d 是星期几，0 表示星期一，1 表示星期二，以此类推。

示例：日期 2021-06-06 16:42:45 是星期几。

```
mysql> select weekday("2021-06-06 16:42:45");
+--------------------------------+
| weekday("2021-06-06 16:42:45") |
+--------------------------------+
|                              6 |
+--------------------------------+
1 row in set (0.00 sec)
```

A.3.48　year(d)：返回指定日期的年份

返回日期 d 的年份。

示例：返回 2021-06-06 16:42:45 的年份。

```
mysql> select year("2021-06-06 16:42:45");
+-----------------------------+
| year("2021-06-06 16:42:45") |
+-----------------------------+
|                        2021 |
+-----------------------------+
1 row in set (0.00 sec)
```

A.4　MySQL 高级函数

A.4.1　bin(x)：返回字符串的二进制编码

返回 x 的二进制编码。

示例：返回 255 的二进制编码。

```
mysql> select bin(255);
+----------+
| bin(255) |
+----------+
| 11111111 |
+----------+
1 row in set (0.00 sec)
```

A.4.2　binary(s)：将字符串转换为二进制数

将字符串 s 转换为二进制数。

示例：将字符串"clay"转换为二进制数。

```
mysql> select binary "clay";
+-------------------------------+
| binary "clay"                 |
+-------------------------------+
| 0x636C6179                    |
+-------------------------------+
1 row in set (0.00 sec)
```

A.4.3　case expression：表达式分支

```
case expression
   when condition1 then result1
   when condition2 then result2
   ...
   when conditionN then resultN
   else result
end
```

case 表示函数开始，end 表示函数结束。如果 condition1（条件 1）成立，则返回 result1；如果 condition2 成立，则返回 result2；如果全部不成立，则返回 result；只要其中一个条件成立，其他条件后面的语句就都不执行了。

示例：根据 name 分组，统计出次数，并输出其等级。

```
mysql> select name,(case when singin_count<5 then '<5' when singin_count>=5 and singin_count<10 then '<10' else '>10' end ) as level
    -> from (select name, sum(singin) as singin_count from employee_tbl group by name )t;
+------+-------+
| name | level |
+------+-------+
| Clay | <10   |
| Alsi | <10   |
| Tom  | <5    |
+------+-------+
3 rows in set (0.00 sec)
```

A.4.4　cast(x as type)：数据类型转换

转换数据类型。

示例：将字符串"20210607"转换为日期。

```
mysql> select cast("20210607" as date);
+--------------------------+
| cast("20210607" as date) |
+--------------------------+
| 2021-06-07               |
+--------------------------+
1 row in set (0.00 sec)
```

A.4.5　coalesce(expr1,…,exprn)：返回第一个非空表达式

返回参数中的第一个非空表达式（从左向右）。

示例：返回 null、null、"clay"、null、"clay2"中的第一个非空表达式。

```
mysql> select coalesce(null, null, "clay", null, "clay2");
+---------------------------------------------+
| coalesce(null, null, "clay", null, "clay2") |
+---------------------------------------------+
| clay                                        |
+---------------------------------------------+
1 row in set (0.00 sec)
```

A.4.6　connection_id()：返回唯一连接 ID

返回唯一的连接 ID。

示例：返回唯一的连接 ID。

```
mysql> select connection_id();
+-----------------+
| connection_id() |
+-----------------+
|           12422 |
+-----------------+
1 row in set (0.00 sec)
```

A.4.7　conv(x,f1,f2)：数据进制转换

把 f1 进制数转换为 f2 进制数。

示例：十进制数 255 转换成二进制数。

```
mysql> select conv(255, 10, 2);
+------------------+
| conv(255, 10, 2) |
```

```
+----------------+
| 11111111       |
+----------------+
1 row in set (0.00 sec)
```

A.4.8　convert(s using cs)：求字符串的字符集

求字符串 s 的字符集。

示例：求字符串"a12"的字符集。

```
mysql> select charset('a12');
+----------------+
| charset('a12') |
+----------------+
| utf8mb3        |
+----------------+
1 row in set (0.00 sec)
mysql> select charset(convert('a12' using gbk));
+-----------------------------------+
| charset(convert('a12' using gbk)) |
+-----------------------------------+
| gbk                               |
+-----------------------------------+
1 row in set (0.00 sec)
```

A.4.9　current_user()：返回当前的用户名

返回当前的用户名。

示例：返回当前登录 MySQL 服务器的用户名。

```
mysql> select current_user();
+----------------+
| current_user() |
+----------------+
| root@localhost |
+----------------+
1 row in set (0.00 sec)
```

A.4.10　database()：返回当前的数据库名

返回当前的数据库名。

示例：返回当前进行操作的数据库名。

```
mysql> use DEMO;
Database changed
mysql> select database();
+------------+
| database() |
+------------+
```

```
| DEMO          |
+---------------+
1 row in set (0.00 sec)
```

A.4.11　if(expr,v1,v2)：表达式判断

如果表达式 expr 成立，则返回结果 v1；否则返回结果 v2。

示例：返回 1>0 的结果。

```
mysql> select if(1 > 0,'yes','no');
+----------------------+
| if(1 > 0,'yes','no') |
+----------------------+
| yes                  |
+----------------------+
1 row in set (0.00 sec)
```

A.4.12　ifnull(v1,v2)：null 替换

如果 v1 的值不为 null，则返回 v1；否则返回 v2。

示例：返回 ifnull(null, "clay")的结果。

```
mysql> select ifnull(null,"clay");
+---------------------+
| ifnull(null,"clay") |
+---------------------+
| clay                |
+---------------------+
1 row in set (0.00 sec)
```

A.4.13　isnull(expression)：判断表达式是否为 null

判断表达式是否为 null。

示例 1：判断 isnull("clay")是否为 null。

```
mysql> select isnull("clay");
+----------------+
| isnull("clay") |
+----------------+
|              0 |
+----------------+
1 row in set (0.00 sec)
```

示例 2：判断 isnull(null)是否为 null。

```
mysql> select isnull(null);
+--------------+
| isnull(null) |
+--------------+
```

```
|              1 |
+----------------+
1 row in set (0.00 sec)
```

A.4.14　nullif(expr1, expr2)：字符串相等则返回 null

比较两个字符串，如果字符串 expr1 与 expr2 相等，则返回 null；否则返回 expr1。

示例 1：比较字符串 11 和字符串 1。

```
mysql> select nullif("11","2");
+------------------+
| nullif("11","2") |
+------------------+
| 11               |
+------------------+
1 row in set (0.00 sec)
```

示例 2：比较字符串"a"和字符串"a"。

```
mysql> select nullif("a","a");
+-----------------+
| nullif("a","a") |
+-----------------+
| null            |
+-----------------+
1 row in set (0.00 sec)
```

A.4.15　last_insert_id()：返回最近生成的自增 ID

返回最近生成的自增 ID（auto_increment 值）。

示例：返回最近生成的自增 ID。

```
mysql> drop table if exists userinfo;
Query OK, 0 rows affected (0.01 sec)
mysql> create table userinfo(id int(10) not null auto_increment primary key,
    -> name varchar(20),
    -> age int(2));
Query OK, 0 rows affected, 2 warnings (0.02 sec)
mysql>  insert into userinfo (name,age) values('zhangsan',26);
Query OK, 1 row affected (0.01 sec)
mysql>  insert into userinfo (name,age) values('lisi',26);
Query OK, 1 row affected (0.01 sec)
mysql> select id from userinfo;
+----+
| id |
+----+
|  1 |
|  2 |
+----+
2 rows in set (0.00 sec)
mysql> select last_insert_id();
```

```
+-----------------+
| last_insert_id() |
+-----------------+
|               2 |
+-----------------+
1 row in set (0.00 sec)
```

A.4.16　session_user()：返回当前会话的用户名

返回当前会话的用户名。

示例：返回当前会话的用户名。

```
mysql> select session_user();
+----------------+
| session_user() |
+----------------+
| root@localhost |
+----------------+
1 row in set (0.00 sec)
```

A.4.17　version()：返回数据库的版本号

返回数据库的版本号。

示例：返回当前安装的数据库的版本号。

```
mysql> select version();
+-----------+
| version() |
+-----------+
| 8.0.25    |
+-----------+
1 row in set (0.00 sec)
```